SpringerBriefs in Applied Sciences and Technology

SpringerBriefs present concise summaries of cutting-edge research and practical applications across a wide spectrum of fields. Featuring compact volumes of 50 to 125 pages, the series covers a range of content from professional to academic.

Typical publications can be:

- A timely report of state-of-the art methods
- An introduction to or a manual for the application of mathematical or computer techniques
- A bridge between new research results, as published in journal articles
- A snapshot of a hot or emerging topic
- An in-depth case study
- A presentation of core concepts that students must understand in order to make independent contributions

SpringerBriefs are characterized by fast, global electronic dissemination, standard publishing contracts, standardized manuscript preparation and formatting guidelines, and expedited production schedules.

On the one hand, **SpringerBriefs in Applied Sciences and Technology** are devoted to the publication of fundamentals and applications within the different classical engineering disciplines as well as in interdisciplinary fields that recently emerged between these areas. On the other hand, as the boundary separating fundamental research and applied technology is more and more dissolving, this series is particularly open to trans-disciplinary topics between fundamental science and engineering.

Indexed by EI-Compendex, SCOPUS and Springerlink.

Preface

We are pleased to present *Neuromorphic Computing: Transforming Disaster Management and Resilience in Civil Engineering*. This timely book explores the innovative intersection of neuromorphic computing with civil engineering, focusing on enhancing disaster management as communities worldwide face an increase in natural disasters.

Neuromorphic computing, inspired by the brain's neural processes, offers a revolutionary approach to processing and responding to complex environmental data associated with natural disasters. This book delves into how this technology can improve disaster resilience and management, enhancing predictive accuracy, response times, and resource allocation.

Authored by both academic curiosity and a commitment to applying technology for societal benefit, this work discusses the potential and challenges of neuromorphic computing, including technical hurdles, ethical considerations, and the need for an interdisciplinary approach.

Intended for engineers, researchers, policymakers, and anyone interested in technology's role in disaster management, we hope this book serves as both a comprehensive guide and an inspiration for further innovation.

We extend our gratitude to our colleagues, students, and faculty at the University of Botswana for their support in realizing this project.

As you explore the promising advancements in *Neuromorphic Computing: Transforming Disaster Management and Resilience in Civil Engineering*, we invite you to consider the transformative impact these technologies may have on enhancing community safety and resilience.

Gaborone, Botswana

Dr. Ali Akbar Firoozi
Dr. Ali Asghar Firoozi

Conflict of Interest

The authors declare no conflict of interest. This book was written and submitted solely to advance knowledge and understanding within the fields of neuromorphic computing, civil engineering, and disaster management. No funding or sponsorship was received for the research and preparation of this book. Furthermore, the authors have no financial, consultative, institutional, or other relationships that could lead to a conflict of interest regarding the book's content, research, or conclusions. Any product, company, or proprietary names mentioned in this book are included purely for the sake of completeness and do not imply endorsement by the authors or bias against those not mentioned.

Contents

Chapter 1
Introduction

This chapter introduces the innovative field of neuromorphic computing and its transformative potential within civil engineering, particularly in disaster management. Exploring the theoretical foundations laid by pioneers like Carver Mead, it discusses the development of computing systems that emulate the neural structures and processing methods of the human brain. These neuromorphic systems are poised to revolutionize disaster management by enhancing the speed and accuracy of response through advanced predictive modeling and real-time data analysis. The chapter further delves into the integration of these technologies into existing civil infrastructure, presenting both the remarkable advantages and the complex challenges involved. Through this exploration, we aim to showcase how neuromorphic computing can lead to more resilient structures and safer communities in an era of increasing environmental threats.

1.1 Theoretical Foundations of Neuromorphic Computing

The concept of neuromorphic computing finds its roots in the groundbreaking work of Carver Mead in the 1980s, who drew inspiration from the neurobiological processes of the human brain to propose a new class of electronic systems. This innovative approach to computing, distinct from traditional models, seeks to emulate the neural structure of the brain, leveraging its mechanisms for processing information [1].

Carver Mead's vision was fueled by the observation that the human brain operates in a fundamentally different way than the von Neumann architectures typical of the period. While traditional computers process tasks sequentially with separate units for memory and computation, the brain processes information in a highly parallel and integrated manner through networks of neurons and synapses. Each neuron can be thought of as a node in a complex network, simultaneously processing and memory storage unit, capable of forming dynamic links to numerous other neurons.

© The Author(s), under exclusive license to Springer Nature Switzerland AG 2024 1
A. A. Firoozi, *Neuromorphic Computing*,
SpringerBriefs in Applied Sciences and Technology,
https://doi.org/10.1007/978-3-031-65549-4_1

This architecture allows the brain to perform many operations simultaneously with remarkable efficiency and minimal energy [2].

Neuromorphic engineering, thus, aims to design and build artificial neural systems whose architectural and functional principles are borrowed from biological nervous systems. Central to this field is the development of electronic circuits that mimic neural functions such as synaptic connectivity and plasticity—the ability of synapses to strengthen or weaken over time, based on the amount of activity they experience. This not only includes the static design of neural network models but also involves dynamic systems that adapt and learn from their sensory inputs in real-time [3].

The practical realization of neuromorphic computing has involved the development of specific technologies such as silicon neurons and silicon synapses, which are designed to behave like their biological counterparts. These components are used to construct large-scale systems that simulate the way the brain processes information. For instance, these systems use spikes—discrete events that occur at points in time, rather than continuous signals—to encode and process information. This spiking neural network model is a more accurate reflection of how real neural networks operate, allowing the implementation of learning algorithms that can adapt based on incoming data, much like learning in the biological brain [4].

The journey from theoretical conception to practical applications in neuromorphic computing has not been without challenges. One of the main hurdles has been the complexity of designing circuits that not only mimic the brain's operations but do so in a power-efficient manner. The brain's ability to perform complex calculations quickly with extremely low-power consumption (around 20 W) starkly contrasts with the large energy requirements of conventional supercomputers. Advances in materials science, particularly in the development of memristive technologies, have provided new pathways to replicate the efficiency of synaptic transmission and plasticity on a chip [5, 6].

Looking forward, the field of neuromorphic computing promises to revolutionize not just computing architectures but also the way we understand and interface with various forms of intelligence, both artificial and biological. As researchers continue to unravel the complexities of the brain's architecture and translate these insights into engineered systems, we can anticipate a new era of machines that understand and interact with the world in fundamentally novel ways. This progression toward more sophisticated and autonomous computing systems holds the potential to transform industries and societies in ways we are just beginning to imagine.

1.2 Neuromorphic Computing in Civil Engineering

The intersection of neuromorphic computing and civil engineering represents a frontier of potential, especially in enhancing the capabilities for disaster management. Civil engineering fundamentally involves the application of scientific principles to the design, construction, and maintenance of the built environment. Integrating neuromorphic computing into this discipline facilitates a leap from conventional static design and maintenance strategies to dynamic, adaptive systems that can respond to changing conditions in real-time.

One of the most critical applications of neuromorphic computing in civil engineering is in structural health monitoring (SHM). SHM systems equipped with neuromorphic computing capabilities can process and analyze data from a myriad of sensors embedded in infrastructure, such as accelerometers, strain gauges, and temperature sensors, among others. These sensors continuously collect data on various aspects of a structure's health, such as vibrations, deformations, and stresses, which are crucial for assessing its integrity [7].

Neuromorphic systems can decipher this complex data stream in real-time, identifying patterns that precede structural failures. For example, slight changes in vibration patterns could indicate a shift in structural load bearing, potentially flagging an imminent risk long before traditional monitoring systems would. By enabling earlier interventions, such systems significantly enhance the safety and longevity of infrastructure, reducing the risk of catastrophic failures [8].

Beyond monitoring, neuromorphic computing can revolutionize the way civil engineers approach construction and maintenance. Traditional methods often rely on scheduled inspections and maintenance, which can be both costly and ineffective if the timing does not align with the emergence of issues. Neuromorphic systems introduce the capability for predictive maintenance, where decisions about when to repair or replace components are driven by actual needs as indicated by real-time data, rather than pre-set schedules.

This shift not only helps in allocating resources more efficiently but also reduces downtime and extends the lifespan of infrastructure components. Furthermore, these systems can aid in the construction phase by optimizing designs based on predictive models that account for various factors like load conditions, weather impact, and material degradation over time [9].

In the context of disaster management, neuromorphic computing can play a pivotal role in both preparedness and response. By integrating neuromorphic systems with environmental monitoring technologies, engineers can develop more accurate predictive models for natural disasters such as earthquakes, floods, and hurricanes. These models can forecast disaster impacts on structures and infrastructures, enabling preemptive evacuations and preparations that can save lives and reduce economic losses.

During a disaster, real-time data processing becomes critical. Neuromorphic computing can enhance the responsiveness of emergency management systems by quickly analyzing data from multiple sources, including satellite imagery, ground sensor networks, and social media feeds, to provide a comprehensive assessment of the situation. This information can be crucial for directing rescue and relief efforts more effectively, ensuring that resources are deployed where they are needed most [10, 11].

The future of neuromorphic computing in civil engineering looks toward not only addressing current challenges but also opening new avenues for innovation. As technology matures, it could lead to the development of fully autonomous systems capable of self-repair and adaptation to environmental changes. Such advancements could significantly reduce human exposure to dangerous conditions, particularly in post-disaster scenarios, by automating risky assessments and repairs.

1.3 Disaster Management Applications

Disaster management, crucial for mitigating the effects of natural catastrophes, stands to be gained substantially from the integration of neuromorphic computing. The traditional frameworks for managing disasters are often hampered by slow data processing, which can delay crucial decisions. Neuromorphic computing, with its superior speed and efficiency, promises to overcome these bottlenecks by enabling more rapid and precise analysis of complex, multisource data.

One of the standout features of neuromorphic computing is its ability to process data in real-time. This is particularly valuable in the context of disaster management, where timely information is critical. For instance, during an earthquake, neuromorphic systems can quickly analyze data from seismic sensors to assess the intensity and epicenter of the quake, providing immediate feedback that can be used to trigger automated emergency responses and alert systems [12].

Neuromorphic computing can also revolutionize the predictive capabilities of disaster management systems. By integrating data from satellite imagery, meteorological stations, and historical disaster databases, neuromorphic systems can help develop advanced models that predict the likelihood, timing, and impact of potential disasters. For hurricanes, this might mean better forecasting of storm paths and intensities, allowing for more effective evacuations and preparation measures. Similarly, for flood management, these systems can analyze weather patterns and hydrological data to forecast flooding events with greater accuracy [13].

Beyond just faster processing and improved predictions, neuromorphic computing enables adaptive learning capabilities that traditional systems lack. This means that disaster response systems can continuously learn from each event, adjusting their models and responses based on new data. Such learning mechanisms can significantly enhance the resilience of disaster management systems by refining their accuracy and efficiency over time, ensuring that with each new disaster, the system becomes more adept at managing and mitigating risks [14].

The rapid processing and adaptive learning capabilities of neuromorphic computing can also improve the coordination of emergency services during disasters. By providing real-time, reliable data to decision-makers, these systems can help optimize the allocation of resources, guide evacuation strategies, and coordinate rescue operations effectively. This ensures that emergency services are not just faster but also more targeted, which can be crucial in reducing the human and material costs of disasters.

Several case studies highlight the practical benefits of applying neuromorphic computing in disaster management. For instance, in regions prone to wildfires, neuromorphic systems have been used to analyze data from thermal sensors and satellites to predict fire spread patterns and guide firefighting efforts efficiently. Similarly, in areas vulnerable to tsunamis, these systems analyze seismic data in conjunction with historical data and oceanographic sensors to provide early warnings that can save lives [15].

Looking forward, the application of neuromorphic computing in disaster management is poised to expand, with new technologies enhancing the speed and accuracy of these systems. The development of more sophisticated sensors, coupled with advances in neuromorphic chip technology, will likely lead to even more powerful applications that could transform how societies prepare for and respond to natural disasters.

1.4 Technical Challenges and Ethical Considerations

The implementation of neuromorphic computing in disaster management, though highly beneficial, comes with its set of technical and ethical challenges that must be navigated carefully. These challenges are not just hurdles to be overcome, but also opportunities to refine the technology and its applications to better serve society while respecting individual rights and ensuring security.

Integration with existing systems: One of the foremost technical challenges is the integration of neuromorphic computing systems with existing infrastructure. Civil engineering infrastructures are often composed of legacy systems that were not designed to interact with advanced computing technologies. Upgrading these systems to be compatible with neuromorphic computing requires significant redesign and retrofitting, which can be costly and time-consuming [16].

Complexity in design and maintenance: Neuromorphic systems are inherently complex due to their emulation of neural processes. Designing these systems requires a deep understanding of both neuroscience and advanced engineering principles. Moreover, maintaining such systems demands ongoing support from specialists who can manage and troubleshoot high-tech components that ordinary IT staff may not be familiar with, thereby increasing operational costs and complexity [17].

Scalability issues: Deploying neuromorphic computing solutions on a large scale, necessary for widespread disaster management applications, poses significant challenges. Scalability issues can arise from the need for extensive sensor networks, high data throughput, and real-time processing capabilities that must be managed and maintained across large geographic areas and varied environmental conditions [18].

Data privacy: The use of neuromorphic computing in disaster management involves the collection, processing, and storage of vast amounts of data, some of which can be highly sensitive. Ensuring the privacy of this data is paramount, as leaks or unauthorized access could lead to significant breaches of privacy. This is particularly critical when dealing with data that can be traced back to individuals, such as location data during disasters [19].

Surveillance concerns: The capability of neuromorphic systems to continuously monitor and analyze environmental and structural data can also raise concerns about surveillance. While beneficial for public safety, there is a thin line between neces-sary observation and invasive surveillance. Balancing these aspects is crucial to maintaining public trust and ensuring that such technologies are not misused [20].

Dependency on automated systems: There is also the risk of over-dependence on automated systems. In disaster management, critical decisions need to be made swiftly and accurately, and while neuromorphic computing can aid significantly in this regard, over-reliance on automation could lead to complacency in human over-sight. Ensuring that human decision-makers remain in the loop and are not entirely replaced by automated processes is important to managing unexpected situations and ethical dilemmas effectively [21].

To address these technical and ethical challenges, ongoing research and devel-opment are essential. Investing in robust cybersecurity measures, designing systems with built-in privacy protections, and creating clear protocols for human oversight can help mitigate many of the risks associated with neuromorphic computing in disaster management. Furthermore, engaging with stakeholders, including the public and ethical boards, can provide diverse insights that help shape the development of these technologies in a socially responsible manner.

1.5   Vision for the Future

The trajectory of neuromorphic computing within the field of civil engineering suggests a transformative future, particularly in enhancing resilience against natural disasters. As technological advancements continue to drive improvements in neuro-morphic hardware and software, the integration of these systems into civil engi-neering practices is poised to become more prevalent and impactful.

Future developments in neuromorphic computing are expected to significantly enhance the ability of civil infrastructures to withstand and adapt to natural disasters. By improving the accuracy and speed of data analysis, these systems can provide early warnings for earthquakes, floods, hurricanes, and other catastrophic events with greater precision. This capability will be critical in preemptively evacuating areas, deploying emergency services more effectively, and managing resources efficiently, thereby saving lives and reducing the economic fallout from disasters [22].

As neuromorphic technologies mature, a key focus will be on reducing the costs associated with these systems. Advances in manufacturing processes, mate-rials science, and energy efficiency are expected to lower the financial barriers to implementing neuromorphic systems, making them more accessible to a broader range of applications within civil engineering. This increased accessibility could revolutionize how smaller municipalities and developing regions approach disaster management, potentially leveling the playing field in terms of disaster preparedness and response capabilities [23].

Looking ahead, neuromorphic computing is set to become a staple in mainstream civil engineering applications. This integration will likely be facilitated by the devel-opment of standardized tools and platforms that allow engineers to easily implement and manage neuromorphic systems. As civil engineers become more familiar with these technologies, their potential applications will expand beyond disaster management to include other areas such as traffic flow optimization, environmental monitoring, and smart city initiatives, further enhancing the efficiency and sustainability of urban environments [24].

As neuromorphic computing technologies evolve, so too will the ethical and regulatory landscapes that govern their use. It will be crucial for policymakers, engineers, and the public to engage in ongoing dialogue about the implications of these technologies. Developing comprehensive guidelines and regulations to manage the ethical challenges discussed earlier will be key to ensuring that neuromorphic computing is used responsibly and beneficially within civil engineering [25].

The future of neuromorphic computing in civil engineering will also depend on continued collaborative and interdisciplinary research. This collaboration will not only involve engineers and computer scientists but also urban planners, environmental scientists, and community stakeholders. Such a multidisciplinary approach will ensure that the development of neuromorphic technologies aligns with broader societal goals and addresses the real-world challenges faced by communities around the globe.

1.6 Conclusion

This introductory chapter sets the stage for a deeper exploration into the specifics of neuromorphic computing applications in disaster management within civil engineering. It outlines the technological foundations, practical applications, and the broader implications of adopting such advanced systems in the context of increasing global disaster risks. As the subsequent sections will detail, the integration of neuromorphic computing into disaster management strategies offers a forward-looking approach to enhancing infrastructure resilience and reducing vulnerabilities in our increasingly complex and interconnected world.

References

1. E.A. Vittoz, Analog VLSI implementation of neural networks, in *Handbook of Neural Computation* (CRC Press, 2020), pp. E1-3
2. J.H. Wijekoon, P. Dudek, Compact silicon neuron circuit with spiking and bursting behaviour. Neural Netw. **21**(2–3), 524–534 (2008). https://doi.org/10.1016/j.neunet.2007.12.037
3. G. Indiveri, B. Linares-Barranco, T.J. Hamilton, A.V. Schaik, R. Etienne-Cummings, T. Delbruck, K. Boahen et al., Neuromorphic silicon neuron circuits. Front. Neurosci. **5**, 73 (2011). https://doi.org/10.3389/fnins.2011.00073
4. S.B. Furber, F. Galluppi, S. Temple, L.A. Plana, The spinnaker project. Proc. IEEE **102**(5), 652–665 (2014). https://doi.org/10.1109/JPROC.2014.2304638
5. L. Chua, Resistance switching memories are memristors, in *Handbook of Memristor Networks* (2019), pp. 197–230. https://doi.org/10.1007/978-3-319-76375-0_6

6. H. Markram, E. Muller, S. Ramaswamy, M.W. Reimann, M. Abdellah, C.A. Sanchez, F. Schür-mann et al., Reconstruction and simulation of neocortical microcircuitry. Cell **163**(2), 456–492 (2015). https://doi.org/10.1016/j.cell.2015.09.029

7. G.V. Joseph, V. Pakrashi, Spiking neural networks for structural health monitoring. Sensors **22**(23), 9245 (2022). https://doi.org/10.3390/s22239245

8. Q. Chen, R. Luley, Q. Wu, M. Bishop, R.W. Linderman, Q. Qiu, AnRAD: a neuromorphic anomaly detection framework for massive concurrent data streams. IEEE Trans. Neural Netw. Learn. Syst. **29**(5), 1622–1636 (2017). https://doi.org/10.1109/TNNLS.2017.2676110

9. N. Sakib, T. Wuest, Challenges and opportunities of condition-based predictive maintenance: a review. Procedia CIRP **78**, 267–272 (2018). https://doi.org/10.1016/j.procir.2018.08.318

10. W. Sun, P. Bocchini, B.D. Davison, Applications of artificial intelligence for disaster management. Nat. Hazards **103**(3), 2631–2689 (2020). https://doi.org/10.1007/s11069-020-04124-3

11. S.N. Aspragkathos, E. Ntouros, G.C. Karras, B. Linares-Barranco, T. Serrano-Gotarredona, K.J. Kyriakopoulos, An event-based tracking control framework for multirotor aerial vehicles using a dynamic vision sensor and neuromorphic hardware, in *2023 IEEE/RSJ International Conference on Intelligent Robots and Systems (IROS)* (IEEE, 2023, October), pp. 6349–6355. https://doi.org/10.1109/IROS55552.2023.10342437

12. N. Kasabov, N.M. Scott, E. Tu, S. Marks, N. Sengupta, E. Capecci, J. Yang et al., Evolving spatio-temporal data machines based on the NeuCube neuromorphic framework: design methodology and selected applications. Neural Netw. **78**, 1–14 (2016). https://doi.org/10.1016/j.neunet.2015.09.011

13. M.R. Hashemi, M.L. Spaulding, A. Shaw, H. Farhadi, M. Lewis, An efficient artificial intelligence model for prediction of tropical storm surge. Nat. Hazards **82**, 471–491 (2016). https://doi.org/10.1007/s11069-016-2193-4

14. P.P. Parlevliet, A. Kanaev, C.P. Hung, A. Schweiger, F.D. Gregory, R. Benosman, C.F. Moss et al., Autonomous flying with neuromorphic sensing. Front. Neurosci. **15**, 672161 (2021). https://doi.org/10.3389/fnins.2021.672161

15. S.P.H. Boroujeni, A. Razi, S. Khoshdel, F. Afghah, J.L. Coen, L. O'Neill, K.G. Vamvoudakis, et al., A comprehensive survey of research towards AI-enabled unmanned aerial systems in pre-, active-, and post-wildfire management. Inf. Fus. 102369 (2024). https://doi.org/10.1016/j.inffus.2024.102369

16. B. Shang, Y. Yi, L. Liu, Computing over space-air-ground integrated networks: challenges and opportunities. IEEE Netw. **35**(4), 302–309 (2021). https://doi.org/10.1109/MNET.011.2000567

17. G.W. Burr, R.M. Shelby, A. Sebastian, S. Kim, S. Kim, S. Sidler, Y. Leblebici et al., Neuromorphic computing using non-volatile memory. Adv. Phys.: X **2**(1), 89–124 (2017). https://doi.org/10.1080/23746149.2016.1259585

18. A.R. Young, M.E. Dean, J.S. Plank, G.S. Rose, A review of spiking neuromorphic hardware communication systems. IEEE Access **7**, 135606–135620 (2019). https://doi.org/10.1109/ACCESS.2019.2941772

19. Y. Pi, N.D. Nath, A.H. Behzadan, Convolutional neural networks for object detection in aerial imagery for disaster response and recovery. Adv. Eng. Inform. **43**, 101009 (2020). https://doi.org/10.1016/j.aei.2019.101009

20. T.D. Räty, Survey on contemporary remote surveillance systems for public safety. IEEE Trans. Syst. Man Cybern. Part C (Appl. Rev.) **40**(5), 493–515 (2010). https://doi.org/10.1109/TSMCC.2010.2042446

21. Y. Senarath, R. Pandey, S. Peterson, H. Purohit, Citizen-helper system for human-centered AI use in disaster management, in *International Handbook of Disaster Research* (Springer Nature Singapore, Singapore, 2023), pp. 477–497. https://doi.org/10.1007/978-981-19-8388-7_34

22. A. Sampathkumar, M. Tesfayohani, S.K. Shandilya, S.B. Goyal, S.S. Jamal, P.K. Shukla, M. Albeedan, et al., Internet of Medical Things (IoMT) and reflective belief design-based big data analytics with Convolution Neural Network-Metaheuristic Optimization Procedure (CNN-MOP), in *Computational Intelligence and Neuroscience* (2022). https://doi.org/10.1155/2022/2898061

23. M. Hu, H. Li, Y. Chen, Q. Wu, G.S. Rose, R.W. Linderman, Memristor crossbar-based neuro-morphic computing system: a case study. IEEE Trans. Neural Netw. Learn. Syst. **25**(10), 1864–1878 (2014). https://doi.org/10.1109/TNNLS.2013.2296777

24. P. Pradhananga, M. ElZomor, G. Santi Kasabdji, Identifying the challenges to adopting robotics in the US construction industry. J. Constr. Eng. Manag. **147**(5), 05021003 (2021). https://doi.org/10.1061/(ASCE)CO.1943-7862.0002007

25. P. McCullagh, G. Lightbody, J. Zygierewicz, W.G. Kernohan, Ethical challenges associated with the development and deployment of brain computer interface technology. Neuroethics **7**, 109–122 (2014). https://doi.org/10.1007/s12152-013-9188-6

Chapter 2
Theoretical Foundations

This chapter delves into the theoretical underpinnings of neuromorphic computing, tracing its development from conceptual foundations to advanced implementations. It highlights the evolution from traditional computing architectures to systems that emulate the brain's neural architecture and processing style. Significant attention is given to spiking neural networks (SNNs) and resistive memory technologies such as memristors, which are critical in mimicking the brain's synaptic mechanisms. The exploration includes a detailed analysis of the transformative potential of neuromorphic computing in enhancing civil engineering applications, particularly in disaster management. By understanding these foundations, readers will gain insight into how neuromorphic computing can transcend conventional computational limitations, offering new paradigms for handling complex, dynamic data in real-time disaster scenarios. This chapter sets the stage for appreciating the advanced capabilities of neuromorphic systems in predicting and managing natural disasters more effectively, contributing to safer, more resilient infrastructures.

2.1 Neuromorphic Computing

Neuromorphic computing, at its core, is an interdisciplinary field that merges insights from neuroscience, computer science, and electrical engineering to create computing systems that replicate the brain's architecture and computational approach. This endeavor is not merely about increasing computational speed or efficiency but about revolutionizing how machines process information, learn, and adapt. The genesis of neuromorphic computing can be traced back to the visionary work of Carver Mead in the late 1980s. Mead's pioneering research laid the groundwork for a new class of computing by suggesting that analog electronic circuits could be designed to emulate the neural processing mechanisms found in the human brain. This concept was revolutionary, suggesting that computing could move beyond the binary constraints of traditional digital systems to embrace a more analog and dynamic method of information processing, akin to that of biological systems [1, 2].

The development of neuromorphic computing has been characterized by an ongoing effort to transcend the limitations of von Neumann architecture, which separates memory and processing units, leading to inefficiencies known as the "von Neumann bottleneck." Neuromorphic systems, in contrast, integrate memory and processing in a manner that mirrors neural networks, allowing for simultaneous data storage, processing, and transmission across a densely interconnected network of synthetic neurons. This architecture facilitates a form of computation that is inherently parallel, distributed, and capable of adapting in real-time, qualities that are essential for processing the complex, noisy, and dynamic data streams that characterize many real-world environments [3, 4].

From a technological standpoint, significant advancements have been made in the development of neuromorphic chips, such as IBM's TrueNorth and Intel's Loihi. These chips are designed with energy efficiency and parallel processing capabilities in mind, featuring thousands to millions of artificial neurons and synapses capable of operating at a fraction of the power required by conventional processors. TrueNorth, for instance, exemplifies this with its ability to perform complex cognitive tasks, such as pattern recognition and sensory data processing, using less power than a conventional light bulb. Similarly, Loihi has demonstrated remarkable efficiency in learning and adapting to new information in real-time, paving the way for applications that require autonomous adaptation to changing conditions [5, 6].

Current research in neuromorphic computing is pushing the boundaries of what is possible with this technology. Scientists and engineers are exploring the use of advanced materials and novel computing paradigms to further enhance the speed, efficiency, and adaptability of neuromorphic systems. This includes the development of new types of memristive devices, which can mimic the plasticity of biological synapses, allowing neuromorphic systems to learn and evolve from experience. Additionally, efforts are being made to create more sophisticated models of neural networks that can capture the complexity and functionality of the human brain with greater fidelity [7].

The significance of neuromorphic computing extends beyond the technical realm into practical applications that can address some of the most pressing challenges of our time. Its potential to revolutionize fields such as artificial intelligence, robotics, and the Internet of Things (IoT) is profound. In these and other domains, neuromorphic computing offers the promise of creating more intelligent, efficient, and adaptable systems that can operate autonomously in complex and unpredictable environments. The journey from theoretical concept to practical application is ongoing, but the advances made thus far suggest a future in which neuromorphic computing plays a central role in shaping the next generation of technology [3].

Table 2.1 outlines the key differences between von Neumann architecture, the traditional computing model, and neuromorphic computing across various critical parameters. The comparison highlights the advancements in processing speed,

Table 2.1 Comparison of computing architectures

Parameter	Von Neumann architecture	Neuromorphic computing
Processing speed	Limited by the sequential data processing and the "von Neumann bottleneck."	Significantly faster due to parallel processing capabilities and the integration of memory and processing units
Energy efficiency	Higher energy consumption due to separate processing and memory units, requiring data to be moved back and forth	Enhanced energy efficiency owing to localized processing and reduced data movement, mirroring the energy-efficient nature of biological brains
Adaptability	Relatively static, requiring manual updates to adapt to new data or algorithms	Highly adaptable, with the ability to learn and evolve in real-time through mechanisms like synaptic plasticity
Real-time data processing	Challenges in handling real-time data processing due to bottlenecks and sequential processing	Excellently suited for real-time data analysis and decision-making, leveraging the inherent speed and parallelism
Scalability	Scaling involves increasing processor speed or memory size, often leading to increased energy consumption and heat generation	Scales more naturally by adding more neural network nodes, allowing for efficient expansion without proportional increases in energy use
Suitability for complex data	Struggles with the volume, velocity, and variety of data typical in disaster management scenarios	Excellently suited for processing complex, nonlinear data patterns typical in natural disaster predictions and management

energy efficiency, and adaptability that neuromorphic computing offers, underscoring its potential to revolutionize data processing in civil engineering and disaster management.

Figure 2.1 presents a historical timeline of computing advancements, starting from the inception of the von Neumann architecture in the 1940s through to the development of neuromorphic computing systems in the current decade. This visual guide traces the pivotal technological innovations that have contributed to the evolution of computing paradigms, encapsulating how each advancement has built upon its predecessors to lead us to the sophisticated neuromorphic approaches of today.

2.2 Spiking Neural Networks (SNNs)

Spiking neural networks (SNNs) stand at the forefront of neuromorphic computing, representing the most biologically accurate model of neural computation developed to date. Unlike their predecessors, which include both the perceptron-based first-generation and the continuous activation function-based second-generation neural networks, SNNs introduce a temporal dimension to neural activity. This innovation

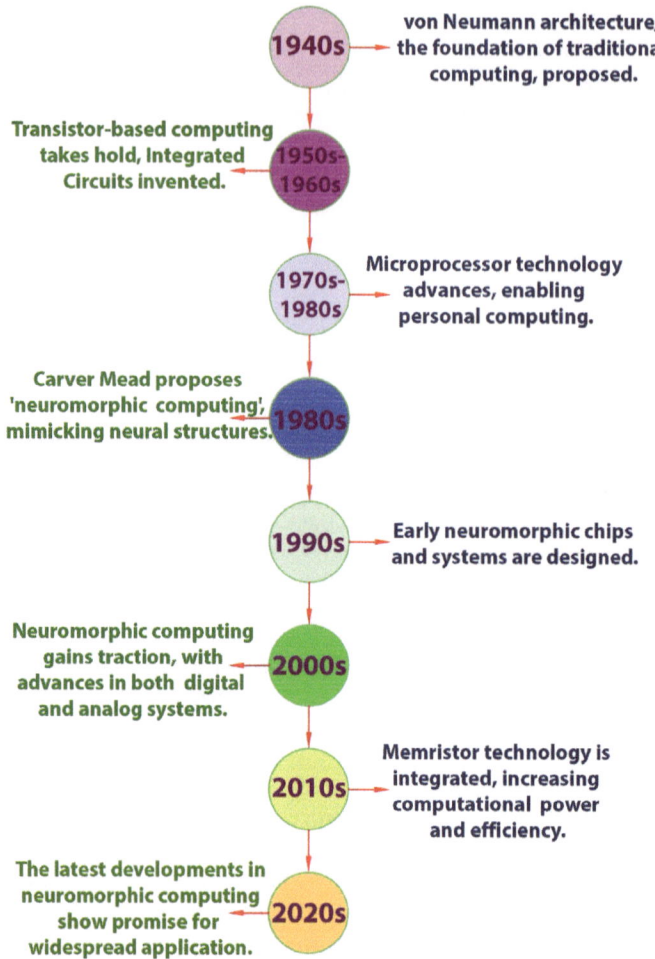

Fig. 2.1 Chronology of computing evolution toward neuromorphic systems

allows them to simulate the precise timing of spikes or action potentials, which are the fundamental means of communication and processing in biological neural networks.

- **The biologically inspired mechanism of SNNs**: The core principle behind SNNs is their ability to mimic the detailed dynamics of biological neurons, including the generation and transmission of spike sequences. Neurons in an SNN communicate by sending spikes to each other, with the timing and frequency of these spikes encoding significant amounts of information. This spike-based communication is a departure from traditional artificial neural networks, which rely on continuous value transmission, offering a richer and more dynamic means of information processing. The incorporation of time as an essential variable enables SNNs to process spatio-temporal patterns in data, making them exceptionally well-suited for tasks that involve sequences, such as speech recognition, natural language processing, and complex sensor integration [8].

- **Advantages of SNNs in computing**: One of the most compelling advantages of SNNs is their efficiency. The event-driven nature of spike processing means that computations are performed only in response to incoming spikes, significantly reducing energy consumption. This contrasts sharply with traditional neural networks, which continuously perform calculations across all nodes, leading to higher computational overheads. Furthermore, the ability of SNNs to encode information in the pattern and timing of spikes allows for a more compact and efficient representation of data, enhancing both computational and energy efficiency [9].

  SNNs also exhibit remarkable capabilities for unsupervised learning, a form of learning where the system develops an understanding of patterns in the data without being provided with explicit labels. This learning is facilitated by mechanisms such as spike timing-dependent plasticity (STDP), a process that adjusts the strength of connections between neurons based on the timing of their spikes. STDP enables SNNs to adapt and learn from the temporal dynamics of incoming data autonomously, mirroring the learning processes observed in the human brain.

- **Implementation and challenges**: Implementing SNNs presents unique challenges, primarily due to their complexity and the need for specialized hardware that can efficiently handle spike-based computations. Early implementations relied on simulations run on conventional computers, which were often inefficient and failed to leverage the full potential of SNNs. However, the development of neuromorphic hardware platforms, such as the TrueNorth and Loihi chips, has provided a more suitable environment for SNNs. These platforms offer parallel, distributed computing architectures and energy-efficient design, aligning with the operational principles of SNNs and enabling their deployment in practical applications [10].

- **The future and potential of SNNs**: The exploration of SNNs is paving the way for advances in various fields, from robotics, where they contribute to the development of more autonomous and adaptive machines, to medicine, where they hold promise for creating better models of brain function and dysfunction. In the realm of artificial intelligence, SNNs offer the potential for creating systems that learn and interact with their environment in more natural and human-like ways, opening new avenues for human–machine interaction and intelligent systems design [11].

Despite the challenges, the ongoing research and development in SNN technology continue to break new ground, bringing us closer to realizing systems that can harness the full power of neuromorphic computing. As we advance, SNNs stand not just as a testament to our growing understanding of the brain but as a beacon guiding us toward a new horizon in computational intelligence.

Equation 2.1 introduces the spike timing-dependent plasticity (STDP) learning rule, a critical mechanism for adjusting synaptic strengths between neurons, pivotal in neuromorphic computing systems. This rule operates on the principle that the timing between presynaptic and postsynaptic spikes determines whether synaptic connections strengthen (potentiation) or weaken (depression). Potentiation occurs when a postsynaptic neuron fires shortly after a presynaptic neuron, suggesting a causal, functional connection that reinforces successful pathways. Conversely, synaptic strength decreases when this firing sequence is reversed, reducing the emphasis on less relevant or unsuccessful pathways. Such dynamic modulation of synaptic efficacy, based on precise spike timing, underlies the ability of neuromorphic systems to learn, adapt, and evolve. It mirrors biological processes of learning and memory formation, enabling these systems to recognize complex patterns and make informed decisions over time, crucial for applications that demand continuous learning from temporal data patterns, such as predictive modeling in disaster management scenarios.

$$\begin{cases} \Delta w = A_+ \exp\left(\frac{-\Delta t}{\tau_+}\right) & \text{for } \Delta t > 0 \\ \Delta w = -A_- \exp\left(\frac{\Delta t}{\tau_-}\right) & \text{for } \Delta t < 0 \end{cases}, \tag{2.1}$$

Δw represents the change in synaptic weight.

Δt is the difference in spike timing between the postsynaptic neuron and the presynaptic neuron, where a positive Δt indicates the post-neuron fires after the pre-neuron and a negative Δt indicates the pre-neuron fires after the post-neuron.

A_+ and A_- are constants that scale the degree of synaptic potentiation and depression, respectively.

τ_+ and τ_- are time constants that determine how quickly the weight change decays over time for potentiation and depression, respectively.

2.3   Resistive Memory Technologies

Resistive memory technologies, integral to the advancement of neuromorphic computing, represent a class of non-volatile memory that holds the key to mimicking the synaptic functionalities of the human brain. Among these, memristors—devices whose resistance changes based on the history of electrical current that has passed through them—stand out due to their ability to emulate the plasticity of biological synapses. This section delves into the operational principles of resistive memory technologies, their significance in neuromorphic computing, and the challenges and prospects associated with their development and implementation. .

- **Operational principles and significance**: The foundational concept of resistive memory technologies lies in their ability to store information in their resistance states, which can be altered through the application of electrical signals. This characteristic makes them particularly suitable for simulating synaptic weights in artificial neural networks, where the strength of synaptic connections needs to be adjusted based on learning algorithms. In biological systems, synaptic plasticity— the ability of synapses to strengthen or weaken over time, based on the level of activity—plays a crucial role in learning and memory. Memristors, with their variable resistance, provide a physical and scalable basis for realizing synaptic plasticity in hardware, offering a path toward creating more brain-like computing systems that learn and adapt [12].

Memristors are not the only form of resistive memory technology; other types include phase-change memory (PCM), conductive-bridging RAM (CBRAM), and ferroelectric RAM (FeRAM). Each of these technologies has unique characteristics and operational mechanisms, but they all share the ability to achieve high-density, low-power, and non-volatile storage. Their incorporation into neuromorphic computing architectures promises to significantly enhance the capacity and efficiency of these systems, enabling the development of devices that can perform complex cognitive tasks with a fraction of the power consumption of traditional computing systems [13].

- **Challenges in development and implementation**: Despite their potential, the integration of resistive memory technologies into neuromorphic computing faces several challenges. One of the primary issues is the variability and reliability of these devices. The resistance states of memristors and other resistive memory devices can be prone to fluctuation due to manufacturing inconsistencies and wear over time, potentially leading to errors in computing. Additionally, the scaling of these technologies to the levels required for practical neuromorphic systems poses significant technical hurdles, including issues related to fabrication, integration with existing complementary metal–oxide–semiconductor (CMOS) technology, and the development of efficient programming and control schemes [14].

  Another challenge is the creation of models and algorithms that can fully exploit the capabilities of resistive memory technologies. While the variable resistance of these devices offers a powerful mechanism for simulating synaptic plasticity, developing algorithms that can effectively utilize this mechanism for learning and memory tasks remains an area of active research. This includes efforts to understand and model the dynamics of memristive devices accurately and to create learning algorithms that can adapt to the specific characteristics and limitations of hardware implementations.

- **Prospects and future directions**: The ongoing research and development in resistive memory technologies are opening new avenues for neuromorphic computing. As scientists and engineers address the challenges associated with these technologies, we are seeing the emergence of more reliable, efficient, and scalable devices that can be integrated into practical computing systems. These advancements hold the promise of revolutionizing various fields, from artificial intelligence and machine learning to robotics and the Internet of Things (IoT), by providing a hardware foundation for creating systems that can learn, adapt, and interact with their environments in ways that mimic the intelligence and efficiency of the human brain [15].

  Furthermore, the exploration of resistive memory technologies is contributing to our understanding of the brain's computational mechanisms. By striving to replicate synaptic plasticity in hardware, researchers are gaining insights into the principles that underlie learning and memory in biological systems. This reciprocal relationship between neuromorphic engineering and neuroscience not only accelerates the development of advanced computing technologies but also enriches our knowledge of the brain's structure and function.

In summary, resistive memory technologies are at the heart of the neuromorphic computing revolution, offering the potential to transcend traditional computing paradigms and usher in a new era of cognitive computing. Despite the challenges that lie ahead, the prospects for these technologies are vast, promising a future where computing systems can learn and think in ways that are currently the domain of biological brains.

Table 2.2 presents a comparison of various resistive memory technologies, including memristors, phase-change memory (PCM), conductive-bridging RAM (CBRAM), and ferroelectric RAM (FeRAM). It highlights their operating principles, advantages, and current limitations, offering insights into their applicability in neuromorphic computing for disaster management.

Table 2.2 Overview of resistive memory technologies and their characteristics

Technology	Principle of operation	Advantages	Current limitations
Memristors	Resistance changes based on the history of voltage and current passed through the device	High density, low-power consumption, and the ability to emulate synaptic plasticity	Variability in performance, challenges in mass production, and long-term durability concerns
Phase-change memory (PCM)	Uses heat from electrical current to switch between amorphous and crystalline states, changing the resistance	Non-volatile, high speed, and good endurance	High programming current, thermal cross-talk, and scalability issues
Conductive-bridging RAM (CBRAM)	Forms and dissolves conductive filaments in an electrolyte to change resistance	Low-power consumption, high switching speed, and excellent scalability	Reliability issues over time, and the need for precise control of filament formation
Ferroelectric RAM (FeRAM)	Utilizes the polarization of a ferroelectric layer to represent data	Non-volatile, low-power usage, and fast read/write cycles	Lower storage density compared to other technologies and complex fabrication processes

2.4 Application in Disaster Management

The intersection of neuromorphic computing and disaster management represents a frontier where advanced computational strategies can significantly enhance the capacity to predict, respond to, and recover from catastrophic events. This fusion is predicated on the unique capabilities of neuromorphic systems, particularly their efficiency in real-time data processing, adaptability through learning algorithms, and proficiency in handling complex, dynamic systems. In the context of disaster management, these attributes translate into potential game-changing applications that can revolutionize how societies prepare for and mitigate the impacts of disasters.

- **Enhancing predictive models**: The core of disaster management lies in the ability to predict and prepare for potential catastrophic events. Neuromorphic computing, with its advanced neural network models like SNNs, offers an unprecedented opportunity to improve the accuracy and timeliness of predictive models for various natural disasters, including earthquakes, floods, and hurricanes. By efficiently processing vast datasets from diverse sources—such as seismic sensors, satellite imagery, and climate models—neuromorphic systems can detect patterns and anomalies that may precede disasters. Unlike traditional computing models that may struggle with the volume and velocity of data, neuromorphic systems can dynamically learn and adapt, providing forecasts that are both more accurate and faster, thereby extending the window for early warnings and preparations [16].
- **Real-time monitoring and decision support**: Disaster management often requires making critical decisions under time constraints and with incomplete information. Neuromorphic computing can play a pivotal role in supporting these decisions by processing real-time data from a network of sensors deployed across vulnerable regions. These could include hydrological sensors for flood monitoring, structural sensors for assessing building integrity during earthquakes, or atmospheric sensors for tracking hurricanes. Neuromorphic systems can analyze this data in real-time, identify patterns indicative of escalating risks, and generate actionable insights for emergency responders. Moreover, the inherent efficiency of neuromorphic computing ensures that these systems can operate continuously, offering constant monitoring and decision support, which is crucial during the acute phases of disaster response [17].
- **Adaptive and resilient infrastructure**: Beyond prediction and response, neuromorphic computing holds promise for creating a smarter, more resilient infrastructure capable of adapting to changing conditions and even mitigating the impacts of disasters. By integrating neuromorphic systems into the control mechanisms of critical infrastructure—such as power grids, water supply networks, and transportation systems—these systems can learn from past events and real-time data to optimize performance and minimize vulnerability to future disasters. For instance, a neuromorphic-controlled power grid could dynamically reroute power to essential services during a storm, or a transportation system could adapt traffic flows to facilitate emergency evacuations efficiently [18].

Table 2.3 Impact of neuromorphic computing on disaster management: applications and outcomes

Disaster type	Application of neuromorphic computing	Impact on response outcomes
Earthquakes	Real-time analysis of seismic data to predict earthquakes with greater accuracy	Enhanced early warning systems provided crucial minutes for evacuation and preparedness, significantly reducing casualties and property damage
Floods	Integration with hydrological sensor networks for dynamic flood prediction and management	Improved accuracy in flood forecasting allowed for timely evacuations and optimized the operation of flood defense infrastructure, mitigating flood impacts
Urban fires	Deployment of sensor networks for early detection and predictive modeling of fire spread	Faster detection and strategic resource deployment minimized fire spread and damage, facilitating quicker emergency responses and reducing urban fire casualties

- **Challenges and future directions**: While the potential applications of neuromorphic computing in disaster management are vast, significant challenges remain. These include the need for extensive sensor networks to collect the necessary data, the development of algorithms capable of interpreting complex environmental signals and ensuring the reliability and robustness of these systems in disaster-prone conditions. Additionally, integrating these advanced technologies into existing disaster management frameworks requires cross-disciplinary collaboration between engineers, computer scientists, emergency responders, and policymakers [19].

    The future of applying neuromorphic computing in disaster management is promising yet it requires concerted efforts to overcome these hurdles. Continued advancements in neuromorphic hardware, algorithms, and system integration will be key to realizing its full potential. Moreover, pilot projects and case studies demonstrating the effectiveness of neuromorphic systems in disaster scenarios will be crucial for gaining the trust and support of stakeholders.

In conclusion, the application of neuromorphic computing to disaster management offers a pathway to significantly enhance our ability to predict, respond to, and recover from disasters. By leveraging the brain-inspired capabilities of neuromorphic systems, disaster management can move toward more proactive, adaptive, and resilient strategies, ultimately saving lives and reducing the economic impact of disasters.

Table 2.3 showcases several real-world applications of neuromorphic computing in the field of disaster management. It highlights specific types of disasters, including earthquakes, floods, and urban fires, and delineates the positive impact that neuromorphic computing has had on improving response outcomes, underscoring the technology's transformative potential.

References

1. S. Srivastava, S.S. Rathod, Silicon neuron-analog CMOS VLSI implementation and analysis at 180 nm, in *2016 3rd International Conference on Devices, Circuits and Systems (ICDCS)* (IEEE, 2016, March), pp. 28–32. https://doi.org/10.1109/ICDCSyst.2016.7570617
2. Y. Ding, Y. Zhang, X. Zhang, P. Chen, Z. Zhang, Y. Yang, Q. Liu et al., Engineering spiking neurons using threshold switching devices for high-efficient neuromorphic computing. Front. Neurosci. **15**, 786694 (2022). https://doi.org/10.3389/fnins.2021.786694
3. S.A. Aamir, Y. Stradmann, P. Müller, C. Pehle, A. Hartel, A. Grübl, K. Meier et al., An accelerated LIF neuronal network array for a large-scale mixed-signal neuromorphic architecture. IEEE Trans. Circuits Syst. I Regul. Pap. **65**(12), 4299–4312 (2018). https://doi.org/10.1109/TCSI.2018.2840718
4. S. Furber, Large-scale neuromorphic computing systems. J. Neural Eng. **13**(5), 051001 (2016). https://doi.org/10.1088/1741-2560/13/5/051001
5. A. Basu, L. Deng, C. Frenkel, X. Zhang, Spiking neural network integrated circuits: a review of trends and future directions, in *2022 IEEE Custom Integrated Circuits Conference (CICC)* (IEEE, 2022, April), pp. 1–8. https://doi.org/10.1109/CICC53496.2022.9772783
6. R.A. Nawrocki, R.M. Voyles, S.E. Shaheen, A mini review of neuromorphic architectures and implementations. IEEE Trans. Electron Devices **63**(10), 3819–3829 (2016). https://doi.org/10.1109/TED.2016.2598413
7. K. Roy, A. Jaiswal, P. Panda, Towards spike-based machine intelligence with neuromorphic computing. Nature **575**(7784), 607–617 (2019). https://doi.org/10.1038/s41586-019-1677-2
8. A. Tavanaei, M. Ghodrati, S.R. Kheradpisheh, T. Masquelier, A. Maida, Deep learning in spiking neural networks. Neural Netw. **111**, 47–63 (2019). https://doi.org/10.1016/j.neunet.2018.12.002
9. K. Yamazaki, V.K. Vo-Ho, D. Bulsara, N. Le, Spiking neural networks and their applications: a review. Brain Sci. **12**(7), 863 (2022). https://doi.org/10.3390/brainsci12070863
10. M. Pfeiffer, T. Pfeil, Deep learning with spiking neurons: opportunities and challenges. Front. Neurosci. **12**, 409662 (2018). https://doi.org/10.3389/fnins.2018.00774
11. J.K. Eshraghian, M. Ward, E.O. Neftci, X. Wang, G. Lenz, G. Dwivedi, W.D. Lu, et al., Training spiking neural networks using lessons from deep learning. Proc. IEEE (2023). https://doi.org/10.1109/JPROC.2023.3308088
12. F. Xue, X. He, Z. Wang, J.R.D. Retamal, Z. Chai, L. Jing, X. Zhang et al., Giant ferroelectric resistance switching controlled by a modulatory terminal for low-power neuromorphic in-memory computing. Adv. Mater. **33**(21), 2008709 (2021). https://doi.org/10.1002/adma.202008709
13. W. Banerjee, S.H. Kim, S. Lee, D. Lee, H. Hwang, An efficient approach based on tuned nanoionics to maximize memory characteristics in Ag-based devices. Adv. Electron. Mater. **7**(4), 2100022 (2021). https://doi.org/10.1002/aelm.202100022
14. X. Wu, V. Saxena, Dendritic-inspired processing enables bio-plausible STDP in compound binary synapses. IEEE Trans. Nanotechnol. **18**, 149–159 (2018). https://doi.org/10.1109/TNANO.2018.2871680
15. S. Yu, Neuro-inspired computing with emerging nonvolatile memorys. Proc. IEEE **106**(2), 260–285 (2018). https://doi.org/10.1109/JPROC.2018.2790840
16. K. Saini, S. Kalra, S.K. Sood, An integrated framework for smart earthquake prediction: IoT, fog, and cloud computing. J. Grid Comput. **20**(2), 17 (2022). https://doi.org/10.1007/s10723-022-09600-7
17. S. Bande, V.V. Shete, Smart flood disaster prediction system using IoT & neural networks, in *2017 International Conference On Smart Technologies For Smart Nation (SmartTechCon)* (IEEE, 2017, August), pp. 189–194. https://doi.org/10.1109/SmartTechCon.2017.8358367
18. J. Li, L. Liu, C. Zhao, K. Hamedani, R. Atat, Y. Yi, Enabling sustainable cyber physical security systems through neuromorphic computing. IEEE Trans. Sustain. Comput. **3**(2), 112–125 (2017). https://doi.org/10.1109/TSUSC.2017.2717807

19. F. Fotovatikhah, M. Herrera, S. Shamshirband, K.W. Chau, S. Faizollahzadeh Ardabili, M.J. Piran, Survey of computational intelligence as basis to big flood management: challenges, research directions and future work. Eng. Appl. Comput. Fluid Mech. **12**(1), 411–437 (2018). https://doi.org/10.1080/19942060.2018.1448896

Chapter 3
Disaster Management in Civil Engineering

3.1 Introduction

This chapter explores the critical role of disaster management within the field of civil engineering, a discipline integral to protecting lives, safeguarding infrastructure, and maintaining community continuity in the face of natural and man-made disasters. Covering the historical development and current practices in disaster management, it emphasizes the multidisciplinary approach required to mitigate risks and enhance the resilience of structures and communities. The chapter outlines the four phases of disaster management—mitigation, preparedness, response, and recovery—detailing strategies and technologies that civil engineers employ to tackle the challenges posed by disasters. It discusses how advancements in technology, especially neuromorphic computing, are revolutionizing this field by enhancing predictive models, improving real-time decision-making, and facilitating the development of smart, adaptable infrastructure. Through this detailed examination, the chapter highlights how civil engineering continues to evolve, driven by innovation and a commitment to public safety, aiming to build stronger, more resilient communities capable of withstanding future disasters.

3.2 Overview of Disaster Management

Disaster management within the civil engineering domain is a complex and critical endeavor, aimed at protecting human lives, infrastructure, and the environment from the adverse effects of disasters. This multidisciplinary approach encompasses a wide range of activities, from the design and construction of resilient structures to the implementation of emergency response strategies. By diving deeper into the traditional approaches to disaster management in civil engineering, we can appreciate

© The Author(s), under exclusive license to Springer Nature Switzerland AG 2024
A. A. Firoozi, *Neuromorphic Computing*,
SpringerBriefs in Applied Sciences and Technology,
https://doi.org/10.1007/978-3-031-65549-4_3

the depth and breadth of efforts involved in mitigating the impact of natural and man-made catastrophes.

- **Mitigation**: Mitigation efforts are foundational to disaster management, focusing on long-term measures to reduce or eliminate the risks associated with disasters. In civil engineering, this involves the strategic planning and design of buildings, bridges, roads, and other infrastructure to withstand potential disasters such as earthquakes, floods, hurricanes, and tsunamis. Techniques such as seismic retrofitting, the construction of levees and floodwalls, and the incorporation of wind-resistant materials are examples of mitigation strategies that engineers deploy. Beyond structural measures, mitigation also includes land use planning practices that discourage development in high-risk areas, such as floodplains or unstable slopes, thereby minimizing potential damage and loss of life [1].
- **Preparedness**: Preparedness in civil engineering revolves around developing plans and capacities to respond effectively to disaster events. This phase involves assessing the vulnerability of infrastructure to various disaster scenarios and devising contingency plans to ensure rapid and efficient emergency response. Civil engineers contribute to preparedness by designing evacuation routes, establishing emergency communication systems, and creating redundancy in critical infrastructure systems to ensure their continued operation during disasters. Training and drills for emergency responders, community disaster education programs, and the stockpiling of necessary supplies and equipment are also key components of preparedness that help communities brace for the impact of potential disasters [2, 3].
- **Response**: The response phase is characterized by immediate actions taken in the aftermath of a disaster to save lives, protect property, and mitigate further damage. Civil engineers play a vital role in emergency response efforts, conducting rapid damage assessments of infrastructure to determine safety and usability, providing technical expertise in search and rescue operations, and facilitating the repair of critical facilities such as roads, bridges, and utilities to restore essential services. Effective disaster response relies on coordination among various stakeholders, including government agencies, emergency services, non-profit organizations, and the affected communities, to ensure that resources are mobilized and deployed where they are most needed [4].
- **Recovery**: Recovery involves the process of restoring the affected community to its pre-disaster state or better, through rebuilding and rehabilitation efforts. For civil engineers, recovery is an opportunity to incorporate disaster risk reduction measures into the reconstruction of infrastructure, applying lessons learned from the disaster to build back better. This phase can encompass a range of activities, from repairing damaged structures and infrastructure to redesigning and reconstructing them with improved resilience against future disasters. Recovery efforts also focus on revitalizing the local economy, restoring environmental conditions, and supporting the psychological and social well-being of the affected population [5, 6].

Fig. 3.1 Disaster management cycle in civil engineering

The integration of these four phases—mitigation, preparedness, response, and recovery—forms a comprehensive approach to disaster management in civil engineering. This approach not only aims to minimize the impact of disasters when they occur but also seeks to build resilient communities capable of withstanding future events. Through diligent planning, innovative design, and effective implementation of disaster management practices, civil engineers play an indispensable role in safeguarding society against the unpredictable forces of nature and human-induced hazards.

Figure 3.1 represents the disaster management cycle in civil engineering, highlighting the continuous efforts to strengthen community resilience. It includes four key phases—risk reduction design, emergency planning, immediate action, and rebuild and improve—with the central goal of building resilience in communities. Each phase is critical to developing a robust approach to disaster management in the field of civil engineering.

Table 3.1 outlines specific engineering strategies and technologies employed in the mitigation of and preparedness for disasters. By categorizing these strategies into mitigation (aimed at reducing disaster risk) and preparedness (focused on planning for an effective response), the table showcases the critical role of civil engineering in enhancing community resilience against natural disasters.

3.3 Challenges

The pursuit of effective disaster management in the realm of civil engineering is fraught with a myriad of challenges. These obstacles span the entire disaster management cycle, from prediction and preparedness to response and recovery, each

Table 3.1 Engineering strategies for disaster mitigation and preparedness

Strategy type	Strategy/technology	Description and application
Mitigation	Seismic retrofitting	Strengthening existing buildings and structures to withstand seismic shocks, significantly reducing earthquake damage potential
	Flood barriers	Construction of levees, floodwalls, and barrier systems to protect vulnerable areas from flooding
	Fire-resistant materials	Use of materials in construction that are resistant to fire, slowing the spread of urban fires
	Wind-resistant design	Designing buildings and structures to endure high wind speeds, minimizing damage from hurricanes and tornadoes
Preparedness	Evacuation route planning	Designing and establishing clear, efficient evacuation routes to ensure rapid and safe exit during emergencies
	Emergency communication systems	Implementing systems that can disseminate warnings and instructions rapidly across different media platforms
	Disaster-resilient infrastructure	Developing infrastructure that maintains functionality in disaster conditions, such as elevated roads in flood-prone areas
	Public education and drills	Conducting community drills and educational programs to enhance public awareness and readiness for disaster situations

presenting unique complexities and requiring innovative solutions. Understanding these challenges in depth is crucial for developing more resilient and adaptive disaster management strategies.

- **Predictive limitations**: One of the most significant hurdles in disaster management is the inherent uncertainty and complexity involved in predicting natural and man-made disasters. Despite advancements in technology and data collection, accurately forecasting the timing, location, and intensity of events such as earthquakes, tsunamis, and industrial accidents remains a daunting task. This unpredictability complicates planning and mitigation efforts, as civil engineers and disaster management professionals must design infrastructure and response strategies that can withstand a wide range of scenarios. The dynamic nature of risk, influenced by factors such as climate change and urbanization, further exacerbates these predictive challenges [7].
- **Data overload and integration**: The digital age has ushered in an era of unprecedented data availability, offering valuable insights for disaster management. However, the sheer volume and variety of data, from satellite imagery and sensor

networks to social media feeds and crowd-sourced information, can overwhelm existing analytical frameworks. Civil engineers and disaster management professionals often grapple with integrating and synthesizing this data to inform real-time decisions. Moreover, the lack of standardized data formats and interoperability between different systems can hinder the effective use of information, delaying response efforts and complicating recovery processes [8].

- **Infrastructure vulnerability**: A significant portion of the world's infrastructure, from transportation networks to utility systems, was designed and constructed under assumptions that may no longer hold due to changing environmental conditions and risk profiles. Aging infrastructure, maintenance deficits, and the challenge of retrofitting facilities to meet current standards pose substantial risks in the event of a disaster. Civil engineers face the daunting task of prioritizing upgrades and ensuring that new constructions adhere to resilience and sustainability principles, often within the constraints of limited budgets and regulatory frameworks [9].

- **Resource allocation and logistics**: Efficiently allocating limited resources before, during, and after a disaster is a complex and critical challenge. Ensuring the timely distribution of emergency supplies, the deployment of response personnel, and the allocation of funds for recovery efforts requires meticulous planning and coordination. The logistical complexities of operating in disaster-affected areas, compounded by damaged infrastructure and communication breakdowns, can impede the effectiveness of response and recovery operations [10].

- **Communication and coordination**: Effective communication and coordination among a diverse array of stakeholders are paramount for successful disaster management. This includes government agencies at various levels, non-profit organizations, private sector entities, and the affected communities themselves. However, differing priorities, capacities, and information asymmetries can lead to fragmented efforts and inefficiencies. Establishing clear channels of communication, fostering partnerships, and developing integrated response frameworks are essential for overcoming these coordination challenges [11].

Addressing these challenges requires a multidisciplinary approach that leverages technological innovations, advances in data analytics, and enhanced collaboration among stakeholders. By confronting these obstacles head-on, civil engineers and disaster management professionals can develop more effective strategies for reducing disaster risks and enhancing the resilience of communities and infrastructure.

Figure 3.2 illustrates the multifaceted challenges in civil engineering disaster management. Central to the diagram, "Core Challenges in Disaster Management" branches out to specific areas, including predictive limitations, data overload and integration, infrastructure vulnerability, resource allocation and logistics, and communication and coordination. Each branch is further detailed with critical subcomponents, providing a comprehensive overview of the obstacles encountered throughout the disaster management cycle.

Table 3.2 highlights how climate change amplifies the challenges faced in disaster management, offering examples of recent events to illustrate the direct consequences

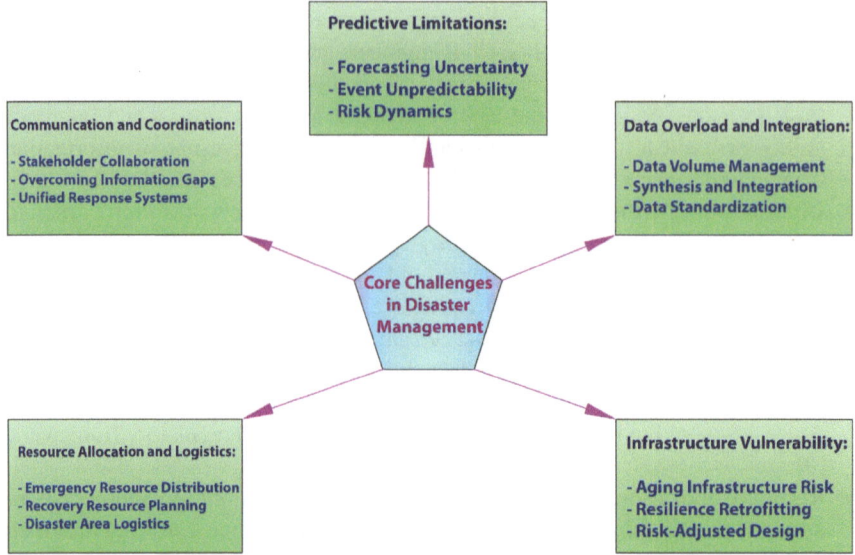

Fig. 3.2 Overview of core challenges in civil engineering disaster management

of these exacerbated challenges. It underscores the urgency of adapting disaster management strategies to account for the changing climate.

3.4 Role of Technology

In the labyrinth of challenges that complicate disaster management within civil engineering, technology emerges not just as a tool but as a pivotal force that can redefine strategies, enhance efficiency, and ultimately save lives. The role of technology, especially advancements like neuromorphic computing, in navigating the complexities of disaster management, is profound and multifaceted. This section delves deeper into how technological innovations are transforming the field, offering solutions to longstanding challenges, and opening new avenues for resilience and adaptation.

- **Advanced-data analytics and predictive modeling**: The advent of big data analytics and machine learning has revolutionized the way data is used in disaster management. By harnessing these technologies, civil engineers can now sift through vast datasets—from weather patterns to geospatial information—in real-time, extracting actionable insights that were previously inaccessible. Predictive modeling, enhanced by artificial intelligence (AI), allows for more accurate forecasting of disasters, enabling preemptive measures to be taken. For instance, AI algorithms can analyze historical data and current conditions to predict the path

Table 3.2 Climate change: exacerbating challenges in disaster management

Challenge exacerbated by climate change	Example event	Consequences and impact on disaster management
Increased frequency and severity of weather-related disasters	Emanuel [12]	These consecutive hurricanes demonstrated the heightened challenge of managing multiple, severe weather events in a single season, stretching emergency response resources thin and highlighting the need for more robust preparedness measures
Rising sea levels and coastal flooding	Vousdoukas et al. [13]	The historic flooding in Venice, exacerbated by sea-level rise, showcased the challenges of protecting heritage sites and urban areas against more frequent and severe coastal flooding, calling for innovative flood defense and urban planning solutions
Intensified wildfires	Abatzoglou and Williams [14]	The unprecedented scale and intensity of the bushfires underlined the increased difficulty in fire prediction, control, and recovery efforts, stressing the importance of enhanced monitoring, early warning systems, and community resilience building
More severe droughts and water scarcity	Swain [15]	The prolonged drought period highlighted the challenges in water resource management, agricultural sustainability, and wildfire risk, emphasizing the need for comprehensive water conservation strategies and drought-resistant infrastructure
Shifts in precipitation patterns leading to floods and landslides	Arnell and Gosling [16]	Severe rainfall led to widespread flooding and landslides across Western Japan, illustrating the complexities in predicting and managing water-related disasters amidst shifting climate patterns, necessitating adaptive water management and land use planning

of hurricanes or the likelihood of urban flooding, facilitating targeted evacuations and the pre-positioning of emergency resources [17].

- **Neuromorphic computing in real-time decision-making**: Neuromorphic computing, with its ability to mimic the human brain's neural architecture, represents a quantum leap in processing speed and efficiency. This technology's inherent capability for parallel processing and its adaptability to changing data patterns make it ideal for real-time decision-making in disaster scenarios. Neuromorphic systems can rapidly analyze inputs from multiple sources, such as sensor networks monitoring structural health or social media platforms providing ground-level updates, to deliver instant assessments and recommendations. This immediacy and accuracy in decision-making are critical during the chaos that typically follows a disaster, helping to streamline response efforts and mitigate further harm [18].
- **Infrastructure monitoring and smart materials**: Technological innovations extend to the monitoring and enhancement of infrastructure resilience. The integration of sensors and smart materials into buildings, bridges, and utility networks

allows for continuous monitoring of structural integrity and environmental conditions. These technologies can detect early signs of distress, such as cracks in a dam or unusual vibrations in a bridge, triggering automatic alerts. Moreover, smart materials equipped with self-healing properties can minimize damage and prolong the lifespan of infrastructure components, reducing vulnerability and enhancing overall resilience [19].

- **Resource optimization and logistics**: Technology also plays a key role in optimizing the allocation of resources and managing the logistics of disaster response. Advanced software platforms can model scenarios, prioritize needs, and orchestrate the distribution of aid, ensuring that resources are deployed efficiently and effectively. Drones and autonomous vehicles, guided by GPS and remote sensing technologies, can navigate disrupted terrains to deliver supplies to isolated areas, conduct aerial surveys, and assist in search and rescue operations, overcoming traditional logistical barriers [20].

- **Enhanced communication and community engagement**: In the digital age, technology facilitates unprecedented levels of communication and community engagement. Mobile applications, social media, and emergency alert systems enable real-time dissemination of information, keeping communities informed and engaged before, during, and after a disaster. Crowdsourced data can provide valuable insights into local conditions, while virtual reality (VR) and augmented reality (AR) technologies can aid in training first responders and educating the public about disaster preparedness [21].

The integration of technology into disaster management strategies represents a paradigm shift, offering hope and tangible solutions to age-old challenges. As these technologies continue to evolve and synergize, they promise to further empower civil engineers and disaster management professionals, enabling them to safeguard lives and infrastructure with unprecedented precision and effectiveness. Equation 3.1 provides a simplified framework for predicting flood risk based on key environmental and hydrological parameters: rainfall intensity, soil saturation levels, and historical river flow rates. By integrating these factors, the model estimates the potential for flooding at any given time, offering a valuable tool for early warning and preparedness efforts. This predictive approach enables more accurate forecasting by dynamically accounting for how current and past conditions influence flood risk, crucial for timely decision-making in flood management strategies.

$$F(t) = R(t) \times S(t) + \int_0^t Q(t')\mathrm{d}t' \qquad (3.1)$$

$F(t)$ represents the flood risk at time t.
$R(t)$ is the rainfall intensity at time t.
$S(t)$ denotes the soil saturation level at time t, indicating how much more water the soil can absorb.
$Q(t')$ represents the river flow rate at a previous time t', integrated over time up to the current moment t to account for accumulated water flow.

Table 3.3 Enhancing disaster response: the impact of advanced technologies

Aspect	Traditional response	Technology-enhanced response
Response time	Often slower due to manual data collection and analysis processes, leading to delayed action	Significantly faster, as automated systems rapidly collect, analyze, and act on data, enabling immediate responses to emerging threats
Decision-making accuracy	Limited by the human capacity to process complex information under pressure, potentially leading to suboptimal decisions	Enhanced by data-driven insights and predictive analytics, leading to more informed and accurate decision-making even in high-stress scenarios
Recovery efforts	Recovery processes can be lengthy and inefficient, hampered by inadequate data on damage extent and resource needs	Accelerated by precise damage assessments and optimized resource allocation, supported by real-time data and AI-driven analysis
Resource allocation	May not be optimized due to a lack of real-time information, resulting in potential misallocation of resources	Dynamic and data-informed, ensuring that resources are directed where they are most needed, enhancing the efficiency and effectiveness of response efforts
Public communication	Often relies on traditional media and may not reach all affected individuals timely, leading to confusion and misinformation	Utilizes multiple digital platforms for widespread, instant communication, ensuring accurate information dissemination and enhancing public safety
Infrastructure resilience	Infrastructure planning and recovery may not fully account for future risk scenarios, and potentially repeating vulnerabilities	Incorporates predictive models and lessons learned into infrastructure rebuilding, focusing on resilience, and reducing future disaster impact

The integral of $Q(t')$ from 0 to t accounts for the historical river flow rates contributing to current water levels.

Table 3.3 contrasts traditional disaster response mechanisms with those augmented by advanced technologies, including neuromorphic computing. It specifically examines differences in response times, decision-making accuracy, and recovery efforts, illustrating the transformative potential of technology in disaster management.

References

1. M. Andrew, Revisiting community-based disaster risk management, in *Reducing Disaster Risks* (Routledge, 2018), pp. 42–52
2. J.C. Gaillard, J.R.D. Cadag, M.M. Rampengan, People's capacities in facing hazards and disasters: an overview. Nat. Hazards **95**, 863–876 (2019). https://doi.org/10.1007/s11069-018-3519-1

3. M.H. Graveline, D. Germain, Disaster risk resilience: conceptual evolution, key issues, and opportunities. Int. J. Disaster Risk Sci. **13**(3), 330–341 (2022). https://doi.org/10.1007/s13753-022-00419-0
4. S.N.G. Gourisetti, M. Mylrea, H. Patangia, Cybersecurity vulnerability mitigation framework through empirical paradigm: enhanced prioritized gap analysis. Futur. Gener. Comput. Syst. **105**, 410–431 (2020). https://doi.org/10.1016/j.future.2019.12.018
5. F.U.R. Khilji, A. Raziq, M. Shoaib, N.S. Baloch, S. Raza, Z. Iqbal, F. Saleem, et al., Expecting the unexpected: nurses' response and preparedness of terrorism-related disaster events in Quetta City, Pakistan. Front. Public Health **9**, 695143 (2021). https://doi.org/10.3389/fpubh.2021.695143
6. D. Sandler, A.K. Schwab, Hazard mitigation and preparedness: an introductory text for emergency management and planning professionals (Routledge, 2021).https://doi.org/10.4324/9781003123897
7. A.V.M. Chand, Place based approach to plan for resilient cities: a local government perspective. Procedia Eng. **212**, 157–164 (2018). https://doi.org/10.1016/j.proeng.2018.01.021
8. Ribot, J., Cause and response: vulnerability and climate in the Anthropocene, in *New Directions in Agrarian Political Economy* (Routledge, 2017), pp. 27–66
9. Cutter, S.L., The perilous nature of food supplies: natural hazards, social vulnerability, and disaster resilience. Environ.: Sci. Policy Sustain. Dev. **59**(1), 4–15 (2017). https://doi.org/10.1080/00139157.2017.1252603
10. L. Smith, Q. Liang, P. James, W. Lin, Assessing the utility of social media as a data source for flood risk management using a real-time modelling framework. J. Flood Risk Manage. **10**(3), 370–380 (2017). https://doi.org/10.1111/jfr3.12154
11. J.R. Elliott, J. Howell, Beyond disasters: a longitudinal analysis of natural hazards' unequal impacts on residential instability. Soc. Forces **95**(3), 1181–1207 (2017). https://doi.org/10.1093/sf/sow086
12. K. Emanuel, Assessing the present and future probability of Hurricane Harvey's rainfall. Proc. Natl. Acad. Sci. **114**(48), 12681–12684 (2017). https://doi.org/10.1073/pnas.1716222114
13. M.I. Vousdoukas, L. Mentaschi, E. Voukouvalas, M. Verlaan, S. Jevrejeva, L.P. Jackson, L. Feyen, Global probabilistic projections of extreme sea levels show intensification of coastal flood hazard. Nat. Commun. **9**(1), 2360 (2018). https://doi.org/10.1038/s41467-018-04692-w
14. J.T. Abatzoglou, A.P. Williams, Impact of anthropogenic climate change on wildfire across western US forests. Proc. Natl. Acad. Sci. **113**(42), 11770–11775 (2016). https://doi.org/10.1073/pnas.1607171113
15. D.L. Swain, A tale of two California droughts: lessons amidst record warmth and dryness in a region of complex physical and human geography. Geophys. Res. Lett. **42**(22), 9999–10010 (2015). https://doi.org/10.1002/2015GL066628
16. N.W. Arnell, S.N. Gosling, The impacts of climate change on river flood risk at the global scale. Clim. Change **134**, 387–401 (2016). https://doi.org/10.1007/s10584-014-1084-5
17. B. Nowell, T. Steelman, A.L.K. Velez, Z. Yang, The structure of effective governance of disaster response networks: insights from the field. Am. Rev. Public Admin. **48**(7), 699–715 (2018). https://doi.org/10.1177/0275074017724225
18. M. Poblet, E. García-Cuesta, P. Casanovas, Crowdsourcing roles, methods and tools for data-intensive disaster management. Inf. Syst. Front. **20**(6), 1363–1379 (2018). https://doi.org/10.1007/s10796-017-9734-6
19. W. Zhang, N. Wang, C. Nicholson, Resilience-based post-disaster recovery strategies for road-bridge networks. Struct. Infrastruct. Eng. **13**(11), 1404–1413 (2017). https://doi.org/10.1080/15732479.2016.1271813
20. L. Palen, A.L. Hughes, Social media in disaster communication, in *Handbook of Disaster Research* (pp. 497–518). https://doi.org/10.1007/978-3-319-63254-4_24
21. M.N. Kamel Boulos, E.M. Geraghty, Geographical tracking and mapping of coronavirus disease COVID-19/severe acute respiratory syndrome coronavirus 2 (SARS-CoV-2) epidemic and associated events around the world: how 21st century GIS technologies are supporting the global fight against outbreaks and epidemics. Int. J. Health Geographics **19**(1), 1–12 (2020). https://doi.org/10.1186/s12942-020-00202-8

Chapter 4
Neuromorphic Computing in Smart Infrastructure

This chapter explores the transformative integration of neuromorphic computing with smart infrastructure within the field of civil engineering, marking a significant evolution from traditional static structures to dynamic, intelligent systems. It delves into how neuromorphic computing enhances the functionality, resilience, and adaptability of urban infrastructures such as transportation systems, energy grids, and buildings. Through advanced data processing capabilities and real-time decision-making, neuromorphic systems enable infrastructure to not only respond to changes in their environment but also predict and adapt to future conditions. The chapter outlines the core aspects of smart infrastructure, discusses the challenges and opportunities presented by integrating neuromorphic computing, and examines case studies where these technologies have been applied in real-world scenarios. This exploration highlights the potential of neuromorphic computing to revolutionize civil engineering practices, making urban areas more efficient, sustainable, and resilient against the challenges posed by rapid urbanization and climate change.

4.1 Smart Infrastructure and Civil Engineering

In the evolving landscape of civil engineering, the concept of smart infrastructure emerges as a cornerstone of modern urban development and sustainability. It signifies a transformative approach to designing, constructing, and managing the built environment, infusing it with digital intelligence to enhance its functionality, resilience, and adaptability. Smart infrastructure encompasses a broad array of elements, from transportation systems and energy grids to water management and building designs, all integrated with cutting-edge technologies like IoT devices, sensors, and advanced materials. This integration facilitates real-time monitoring, analysis, and control, allowing infrastructure to interact with its users and the environment in unprecedented ways.

© The Author(s), under exclusive license to Springer Nature Switzerland AG 2024
A. A. Firoozi, *Neuromorphic Computing*,
SpringerBriefs in Applied Sciences and Technology,
https://doi.org/10.1007/978-3-031-65549-4_4

- **The foundation of smart infrastructure**: At its core, smart infrastructure is designed to address the complex challenges of the twenty-first century, including rapid urbanization, climate change, resource constraints, and the increasing demands of urban populations. By leveraging data and connectivity, smart infrastructure aims to optimize resource use, reduce emissions, improve service delivery, and enhance the quality of life. It is characterized by its ability to collect vast amounts of data through embedded sensors and IoT devices, process this information to gain insights into operational performance, and adapt to changing conditions through automated systems and feedback loops [1].
- **The role of civil engineering**: Civil engineers are at the forefront of creating smart infrastructure, applying their expertise to embed intelligence into the physical fabric of our cities. This involves not only the technical aspects of incorporating sensors and connectivity into infrastructure projects but also the innovative design and planning required to ensure these elements enhance the system's overall performance and sustainability. From the initial stages of conceptualization and design to construction and operation, civil engineers must consider how each component can contribute to a smarter, more integrated infrastructure ecosystem [2].
- **Challenges and opportunities**: The transition to smart infrastructure presents a myriad of challenges, including technical hurdles related to interoperability, scalability, and security. Ensuring the seamless integration of various technologies, protecting against cyber threats, and managing the lifecycle of smart components are critical considerations for civil engineers. Additionally, there are broader societal and ethical issues to address, such as privacy concerns, access to technology, and the potential for increased inequality [3].

  Despite these challenges, the shift toward smart infrastructure offers immense opportunities to reimagine urban environments. It enables more efficient transportation systems that can reduce congestion and pollution, intelligent energy grids that optimize consumption and incorporate renewable sources, and water management systems that enhance resilience to droughts and floods. Furthermore, smart buildings and structures can improve safety and comfort for inhabitants, adapting to their needs in real-time and facilitating a more sustainable relationship between the built environment and the natural world.
- **The future of civil engineering**: As the field of civil engineering continues to evolve, the integration of smart infrastructure becomes increasingly central to its mission. This paradigm shift demands a new set of skills and perspectives, blending traditional engineering principles with expertise in data analytics, cybersecurity, and systems thinking. The future of civil engineering lies in its ability to design and construct infrastructure that is not only physically robust but also digitally intelligent, creating urban spaces that are more livable, resilient, and attuned to the needs of both people and the planet. In this context, smart infrastructure represents not just a technological advancement, but a fundamental rethinking of how infrastructure can serve society in the twenty-first century and beyond [4].

Table 4.1 delineates a range of technologies pivotal to the development of smart infrastructure, highlighting their applications across different civil engineering components. It showcases how IoT devices, advanced materials, and sensors, among others, contribute to enhancing the functionality, resilience, and adaptability of infrastructure systems.

4.2 Integration of Neuromorphic Computing

The integration of neuromorphic computing into smart infrastructure represents a transformative leap forward in civil engineering, embodying a confluence of neuroscience, computing, and infrastructure design. This integration process is not merely about embedding advanced technology into physical structures but about reimagining the very essence of infrastructure to be responsive, adaptive, and efficient.

- **The integration process**: The process of integrating neuromorphic computing systems into infrastructure involves several critical steps, each posing its unique set of challenges and requiring innovative solutions. Initially, it necessitates the deployment of extensive sensor networks capable of capturing a wide array of data, from environmental conditions to structural health and user interactions. These sensors serve as the nervous system of smart infrastructure, collecting the stimuli to which neuromorphic systems will respond [5].

  Following sensor deployment, the next step involves embedding neuromorphic chips or processors within the infrastructure. These chips process data at the edge, i.e., locally where data is collected, enabling real-time analytics and decision-making. This capability is crucial for applications requiring immediate response, such as adaptive traffic control systems in smart cities or real-time flood monitoring and response systems.

  Achieving seamless communication and interoperability between neuromorphic systems and existing digital infrastructure networks is another vital component of the integration process. This ensures that data and insights generated by neuromorphic processors can be shared across different systems and platforms, facilitating coordinated actions and responses across the infrastructure ecosystem.
- **Challenges and solutions**: Integrating neuromorphic computing into infrastructure is fraught with challenges, chief among them being the need for robust data privacy and security measures. The vast amount of data collected and processed by neuromorphic systems could be vulnerable to cyberattacks if not properly secured. Implementing advanced encryption techniques and secure communication protocols is essential to protect data integrity and privacy [6].

  Another challenge lies in the energy consumption and sustainability of neuromorphic systems. While neuromorphic computing is inherently more energy-efficient than traditional computing models, the scale of deployment in infrastructure projects necessitates careful consideration of power sources and energy

Table 4.1 Smart infrastructure technologies and their applications

Technology	Application in civil engineering	Description
IoT devices	Monitoring and control systems for buildings and urban infrastructure	Enable real-time monitoring and control of infrastructure systems, improving efficiency and safety
Advanced materials	Construction of buildings, bridges, and roads	Materials with enhanced properties (e.g., self-healing concrete, ultra-high-performance concrete) that increase durability and reduce maintenance needs
Sensors	Structural health monitoring, environmental monitoring	Collect data on structural integrity, environmental conditions, and other critical parameters, facilitating preventive maintenance and immediate response to issues
Geospatial technologies	Urban planning and disaster risk assessment	Support the analysis and visualization of geographical data for urban planning, risk assessment, and resource management
Renewable energy systems	Energy supply for infrastructure	Incorporate sustainable energy sources, such as solar and wind, into infrastructure, reducing carbon footprint and ensuring energy resilience
Wireless communication networks	Coordination of emergency services, information dissemination	Facilitate rapid communication and coordination among emergency responders and the public during disasters
Artificial intelligence and machine learning	Predictive maintenance, traffic management, and water resource management	Analyze data from various sources to predict maintenance needs, optimize traffic flows, and manage water resources efficiently
Robotics and drones	Inspection and maintenance of hard-to-reach infrastructure components	Perform inspections and maintenance tasks in challenging environments, enhancing safety and efficiency
Digital twins	Simulation and management of infrastructure systems	Create digital replicas of physical infrastructure to simulate performance under various conditions, aiding in optimization and disaster preparedness

use. Solutions include optimizing chip design for lower power consumption and leveraging renewable energy sources to power sensor networks and processors.

Moreover, the technical complexity of neuromorphic systems requires a workforce skilled in interdisciplinary fields, combining knowledge in civil engineering, computer science, and neuroscience. Developing educational and training programs to cultivate such expertise is crucial for the successful integration and maintenance of neuromorphic computing in infrastructure.

- **Case in point**: One of the most compelling applications of neuromorphic computing in smart infrastructure is in enhancing disaster resilience. For instance, neuromorphic systems integrated into flood management infrastructure can analyze data from water level sensors in real-time, using predictive models to anticipate flood events and automatically activate flood defenses or alert communities. Similarly, in earthquake-prone areas, neuromorphic systems can process signals from seismic sensors to detect early warning signs of earthquakes, triggering automatic shutdown procedures for utilities and transportation networks to mitigate damage [7].

The integration of neuromorphic computing into smart infrastructure opens a new frontier in civil engineering, offering the promise of infrastructure that not only supports human activity but actively contributes to the safety, efficiency, and sustainability of urban environments. As this technology matures and scales, its potential to transform our cities and societies is boundless, marking a significant step toward a more resilient and intelligent future.

Equation 4.1 articulates the efficiency of neuromorphic computing in processing real-time data compared to traditional computing methods. Efficiency is quantified as the volume of data processed per unit of time and energy consumed. A higher efficiency ratio (R) signifies a greater efficiency of neuromorphic computing over traditional methods, emphasizing neuromorphic computing's capability to process vast datasets more swiftly and with less energy. This advantage is pivotal in applications requiring real-time analysis and decision-making, such as dynamic disaster response scenarios, where reducing processing time and energy consumption is crucial.

$$\begin{cases} E_N = \frac{D_N}{T_N \times E_{cN}} \\ E_T = \frac{D_T}{T_T \times E_{cT}} \end{cases}, \tag{4.1}$$

where

E_N and E_T are the efficiencies of neuromorphic and traditional computing systems, respectively.

D_N and D_T represent the volume of data processed by neuromorphic and traditional systems, respectively.

T_N and T_T denote the processing time required by neuromorphic and traditional systems, respectively.

E_{cN} and E_{cT} indicate the energy consumption of neuromorphic and traditional systems, respectively.

To compare the efficiency between the two computing systems, we can use the efficiency ratio (R): $R = \frac{E_N}{E_T}$.

The flowchart in Fig. 4.1 delineates the sequential integration process of neuromorphic computing technology within disaster management systems. It begins with "data collection" through sensors and progresses through "data transmission," "neuromorphic processing," and "advanced analysis." The process culminates in "decision support" and "action implementation," with a critical "feedback for learning" loop that informs continuous system improvement and learning.

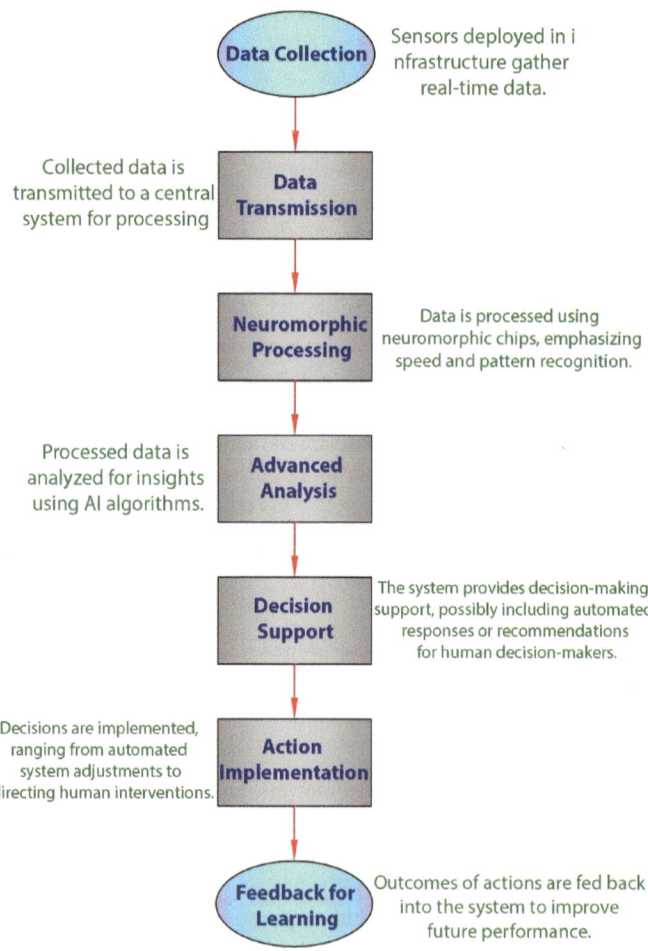

Fig. 4.1 Integration of neuromorphic computing in disaster management

Table 4.2 Navigating the challenges of neuromorphic computing integration

Challenge	Solution or strategy
Scalability and complexity	Develop modular neuromorphic systems that allow for incremental scaling and adapt to various applications without overwhelming complexity
Integration with existing infrastructure	Implement adaptable interfaces and middleware that can translate between neuromorphic computing protocols and those of existing digital infrastructure, ensuring seamless data exchange and system compatibility
Reliability in extreme conditions	Enhance system durability through robust hardware design and protective enclosures; establish redundant systems and fail-safes to maintain operation during critical failures
Data privacy and security	Employ advanced encryption methods and secure data transmission protocols; introduce strict access controls and continuous monitoring to safeguard sensitive information
Energy consumption	Optimize neuromorphic chip design for low-power consumption; integrate energy harvesting technologies and renewable energy sources to power sensor networks and computing units
Technical expertise	Foster interdisciplinary training programs and workshops to equip professionals with the skills required to develop, deploy, and maintain neuromorphic systems
Public perception and acceptance	Conduct outreach and education campaigns to inform the public about the benefits and safeguards of neuromorphic computing in disaster management, addressing concerns transparently to build trust
Regulatory and ethical compliance	Collaborate with regulatory bodies to establish clear guidelines for neuromorphic computing applications; incorporate ethical considerations into the development process to ensure responsible use

Table 4.2 outlines the primary challenges encountered in integrating neuromorphic computing within civil engineering and disaster management frameworks, alongside potential solutions or strategies aimed at addressing these issues. It provides a clear view of the obstacles and the innovative approaches to overcome them, facilitating the successful adoption of neuromorphic computing technologies.

4.3 Case Studies

Exploring the integration of neuromorphic computing within smart infrastructure through case studies illuminates the practical applications and transformative potential of this technology in enhancing disaster management and urban resilience. These detailed scenarios not only demonstrate the current capabilities but also envision the future trajectory of neuromorphic computing in civil engineering.

I. **Earthquake Early Warning System**

- *Context and challenge*: In earthquake-prone regions, the critical challenge is the rapid detection and response to seismic activities to minimize human and structural losses. Traditional seismic monitoring systems often provide limited lead times for evacuation and automated system shutdowns [8].
- *Neuromorphic solution*: A network of neuromorphic computing-based seismic sensors was deployed across a metropolitan area, designed to analyze underground tremors and vibrations in real-time. Leveraging the high-speed processing capabilities of neuromorphic chips, these sensors could differentiate between benign and potentially destructive seismic patterns, providing crucial seconds or even minutes of additional warning time compared to conventional systems.
- *Outcome*: The implementation of this neuromorphic system enabled the automated and immediate shutdown of critical infrastructure, such as gas lines and electrical grids, significantly reducing fire hazards and electrical damage. Public transportation systems received automated signals to halt trains and open subway doors, and emergency response units were pre-alerted to potential impacts, streamlining evacuation and response efforts. The system's success demonstrated the power of neuromorphic computing in providing a more resilient urban environment against earthquakes.

II. **Smart Flood Management in Urban Drainage Systems**

- *Context and challenge*: Urban areas with outdated drainage systems face significant flood risks during heavy rainfall events, exacerbated by climate change and increased surface runoff. Traditional flood management systems struggle with real-time data analysis and predictive modeling, often reacting to flooding events rather than anticipating them [9].
- *Neuromorphic solution*: A city implemented a smart drainage system powered by neuromorphic computing to enhance its flood response strategy. The system integrated sensors throughout the city's drainage network to monitor water levels, flow rates, and weather conditions continuously. Neuromorphic processors analyzed this data in real-time, enabling the system to predict flooding events with high accuracy and dynamically adjust the drainage flow accordingly. This included opening and closing sluice gates and activating pumps, based on predictive models of how rainwater would affect different parts of the city.
- *Outcome*: The smart drainage system significantly mitigated flooding, even in historically vulnerable neighborhoods. By preemptively managing water flow and storage, the city could prevent overflow events and reduce the impact of heavy rainfall. The success of this project highlighted the effectiveness of neuromorphic computing in managing complex, dynamic systems like urban water management infrastructure.

III. **Adaptive Traffic Control for Evacuation Efficiency**

- *Context and challenge*: During disasters, efficiently evacuating people from affected areas is paramount. Traditional traffic management systems often become overwhelmed by sudden increases in demand, leading to congestion and delays when time is of the essence [10].
- *Neuromorphic solution*: An urban center deployed an adaptive traffic control system, utilizing neuromorphic computing to optimize traffic flow during emergency evacuations. By processing real-time data from a network of traffic cameras, vehicle counters, and social media feeds, the neuromorphic system could identify congestion points and predict traffic patterns several steps ahead.
- *Outcome*: In the event of an emergency requiring evacuation, the system dynamically adjusted traffic signals, opened counterflow lanes, and provided real-time routing recommendations to drivers through mobile apps and digital road signs. This proactive management reduced evacuation times by up to 50%, demonstrating the capacity of neuromorphic computing to enhance public safety significantly during critical situations.

These case studies exemplify the profound impact neuromorphic computing can have on smart infrastructure and disaster management. By harnessing the speed, efficiency, and adaptability of neuromorphic systems, civil engineering can leap forward, creating smarter, safer, and more resilient urban environments.

Table 4.3 provides a comparative analysis between traditional disaster management approaches and those augmented by neuromorphic computing. It focuses on key aspects such as prediction accuracy, response times, and recovery efficiency, highlighting the advancements neuromorphic computing brings to the field.

Table 4.3 Traditional versus neuromorphic computing-enhanced disaster management

Aspect	Traditional approach	Neuromorphic computing-enhanced approach
Prediction accuracy	Relies on historical data and statistical models, which may not account for all variables, leading to less accurate predictions	Utilizes real-time data processing and machine learning algorithms, significantly improving the accuracy of disaster predictions by accounting for a wider range of variables
Response times	Can be delayed due to the need for manual data analysis and decision-making processes	Dramatically reduced, as neuromorphic systems can analyze data and make decisions in real-time, facilitating immediate action
Recovery efficiency	Often hampered by the lack of precise data on damage extent and resource needs, leading to slower and less targeted recovery efforts	Enhanced by accurate and timely damage assessments, optimized resource allocation, and adaptive recovery plans, leading to more efficient and effective recovery operations
Resource allocation	May not be optimized due to slower data analysis and communication lag, potentially leading to wastage or misallocation	Improved through dynamic and data-driven resource allocation, ensuring that resources are directed where they are most critically needed on time
Public communication	Relies on traditional communication methods, which may not reach or effectively inform all affected individuals	Employs advanced communication technologies (e.g., social media, emergency apps) for widespread, instant, and interactive public alerts and information dissemination
Infrastructure resilience	Limited by static design parameters and slower incorporation of post-disaster learnings into infrastructure planning	Continuously improved through the integration of real-time data and predictive analytics into infrastructure design and maintenance, increasing resilience against future disasters

References

1. Y. Mehmood, F. Ahmad, I. Yaqoob, A. Adnane, M. Imran, S. Guizani, Internet-of-things-based smart cities: recent advances and challenges. IEEE Commun. Mag. **55**(9), 16–24 (2017). https://doi.org/10.1109/MCOM.2017.1600514
2. M. Ouyang, Critical location identification and vulnerability analysis of interdependent infrastructure systems under spatially localized attacks. Reliab. Eng. Syst. Saf. **154**, 106–116 (2016). https://doi.org/10.1016/j.ress.2016.05.007
3. I.A.T. Hashem, V. Chang, N.B. Anuar, K. Adewole, I. Yaqoob, A. Gani, H. Chiroma, The role of big data in smart city. Int. J. Inf. Manage. **36**(5), 748–758 (2016). https://doi.org/10.1016/j.ijinfomgt.2016.05.002
4. P. Chamoso, A. González-Briones, S. Rodríguez, J.M. Corchado, Tendencies of technologies and platforms in smart cities: a state-of-the-art review. Wirel. Commun. Mobile Comput. (2018). https://doi.org/10.1155/2018/3086854
5. G. Indiveri, S.C. Liu, Memory and information processing in neuromorphic systems. Proc. IEEE **103**(8), 1379–1397 (2015). https://doi.org/10.1109/JPROC.2015.2444094

6. C.D. Schuman, T.E. Potok, R.M. Patton, J.D. Birdwell, M.E. Dean, G.S. Rose, J.S. Plank, A survey of neuromorphic computing and neural networks in hardware (2017). arXiv preprint arXiv:1705.06963. https://doi.org/10.48550/arXiv.1705.06963

7. C.D. James, J.B. Aimone, N.E. Miner, C.M. Vineyard, F.H. Rothganger, K.D. Carlson, S.J. Plimpton, A historical survey of algorithms and hardware architectures for neural-inspired and neuromorphic computing applications. Biol. Inspired Cogn. Arch. **19**, 49–64 (2017). https://doi.org/10.1016/j.bica.2016.11.002

8. K. O'shea, R. Nash, An introduction to convolutional neural networks (2015). arXiv preprint arXiv:1511.08458. https://doi.org/10.48550/arXiv.1511.08458

9. N.K. Upadhyay, H. Jiang, Z. Wang, S. Asapu, Q. Xia, J. Joshua Yang, Emerging memory devices for neuromorphic computing. Adv. Mater. Technol. **4**(4), 1800589 (2019). https://doi.org/10.1002/admt.201800589

10. A. Shrestha, H. Fang, Z. Mei, D.P. Rider, Q. Wu, Q. Qiu, A survey on neuromorphic computing: models and hardware. IEEE Circuits Syst. Mag. **22**(2), 6–35 (2022). https://doi.org/10.1109/MCAS.2022.3166331

Chapter 5
Advanced Predictive Models for Natural Disasters

This chapter delves into the groundbreaking advancements brought by neuromorphic computing to the field of disaster management, particularly in the realm of natural disasters like earthquakes, floods, and urban fires. Through the implementation of neuromorphic computing, predictive models gain unprecedented levels of speed, accuracy, and efficiency, far surpassing the capabilities of traditional computing methods. The chapter explores the specific applications of these advanced models in real-time monitoring and predictive analytics, demonstrating their superiority in responding to natural disaster scenarios. By leveraging the brain-like processing capabilities of neuromorphic systems, these models not only enhance our predictive abilities but also significantly improve the management and mitigation strategies for these catastrophic events. Through comparative analysis with conventional models, the chapter highlights the transformative impact of neuromorphic computing in reshaping disaster preparedness and response, leading to more informed decision-making and ultimately reducing the adverse effects of natural disasters on human life and infrastructure.

5.1 Enhancing Predictive Models for Earthquakes, Floods, and Urban Fires

The advent of neuromorphic computing has ushered in a new era for disaster management, offering sophisticated models that significantly enhance our ability to predict and respond to natural disasters. By leveraging the brain-like processing capabilities of neuromorphic systems, these models promise to outperform traditional computing methods in speed, accuracy, and efficiency. This section explores the innovative applications of neuromorphic computing in predicting and managing earthquakes, floods, and urban fires, culminating in a comparative analysis with conventional models.

A. A. Firoozi, *Neuromorphic Computing*,
SpringerBriefs in Applied Sciences and Technology,
https://doi.org/10.1007/978-3-031-65549-4_5

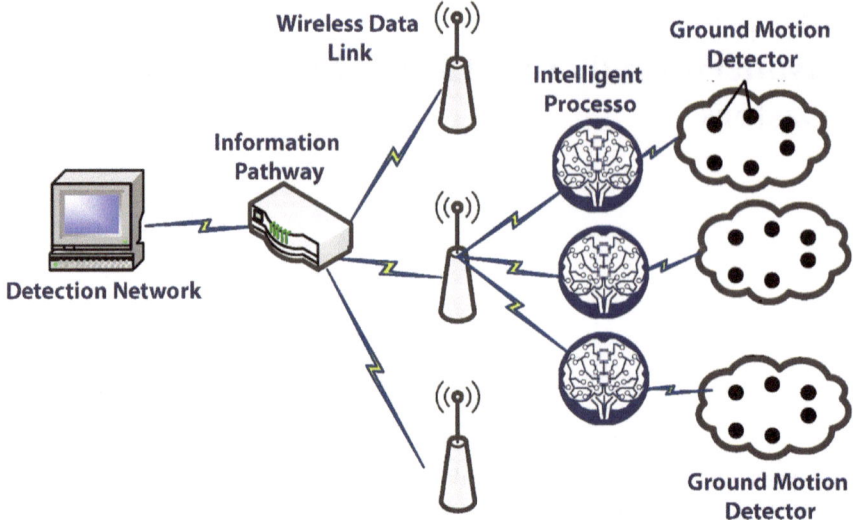

Fig. 5.1 Neuromorphic-based seismic sensor network

Figure 5.1 depicts a sophisticated seismic sensor network utilizing intelligent processors for the rapid analysis of seismic data. Ground motion detectors capture vital seismic activity, which is then transmitted through a wireless data link. The information pathway culminates at the detection network where advanced processing takes place, enabling timely and accurate seismic monitoring and response.

5.2   Earthquakes

The prediction and management of earthquakes represent one of the most challenging yet critical areas in disaster management. The inherently unpredictable nature of seismic events, combined with their potential for catastrophic impact, underscores the urgent need for advanced predictive models. Neuromorphic computing, with its unique capabilities, offers new horizons in earthquake prediction, detection, and response strategies, providing a paradigm shift from traditional methodologies.

- **Development of neuromorphic computing models for earthquake prediction**:
  The development of neuromorphic models for earthquake prediction involves the integration of vast networks of seismic sensors equipped with neuromorphic chips. These sensors are deployed across tectonically active regions, continuously monitoring the Earth's crust for the faintest tremors and vibrations that precede major seismic events. Neuromorphic computing's edge lies in its ability to process and analyze this data in real-time, mimicking the human brain's efficiency and speed [1, 2].

These models employ sophisticated algorithms to interpret the seismic data, learning from historical patterns to identify the precursors of earthquakes. Unlike traditional computing systems, which process data linearly and can be overwhelmed by the sheer volume of information, neuromorphic systems can handle complex, nonlinear data sets with ease. This allows for the detection of subtle anomalies in seismic patterns that may indicate an impending earthquake, offering a potential breakthrough in early warning capabilities.

- **Application in earthquake management**: The application of neuromorphic computing in earthquake management extends beyond prediction. Once a potential seismic event is identified, these systems can quickly analyze potential impact scenarios, considering local geography, urban density, and historical data on building resilience. This analysis enables emergency services to prioritize response efforts, optimize evacuation routes, and deploy resources where they are needed most, even before the earthquake strikes [3].

  Furthermore, neuromorphic models can be integrated with public warning systems, triggering automated alerts to cell phones, broadcast networks, and public alarm systems, providing the public with crucial minutes to take cover or evacuate dangerous areas. This rapid dissemination of warnings is critical in reducing casualties and injuries during seismic events.

- **Comparative advantage**: The comparative advantage of neuromorphic computing models over traditional earthquake prediction methods is stark. Traditional seismic analysis often relies on statistical models that can predict general patterns but lack the precision to provide short-term warnings. These models are also constrained by computational limitations, struggling to process real-time data from multiple sources simultaneously [4].

  Neuromorphic computing, by contrast, excels in real-time data analysis and the identification of complex patterns within vast datasets. Its ability to process information at the edge, near the source of data collection, significantly reduces response times. Moreover, the adaptive learning capabilities of neuromorphic models mean they continually refine their predictive accuracy based on new data, a feature that traditional static models cannot match.

In summary, the advent of neuromorphic computing in earthquake prediction and management heralds a new era in our ability to foresee and mitigate the impacts of seismic events. By harnessing the power of brain-inspired computing, we are on the cusp of significantly advancing our resilience against one of nature's most unpredictable and destructive forces.

Equation 5.1 delineates a neuromorphic model's approach to predicting earthquakes by fusing real-time seismic activity with historical earthquake data. This algorithm employs a logistic function to translate seismic indicators into a probability score, reflecting the likelihood of an impending earthquake. The model's strength lies in its ability to adaptively weigh current observations against established patterns, thereby enhancing prediction accuracy. Through continuous learning from new seismic events, the neuromorphic system refines its coefficients (a, b, and c), improving its predictive capabilities over time and offering vital insights for earthquake preparedness and risk mitigation.

$$P_{eq} = \sigma(a.S + b.H + c), \tag{5.1}$$

where:

$\sigma_{(x)} = \frac{1}{1+e^{-x}}$ is the logistic function, ensuring that the output, P_{eq}, ranges between 0 and 1, representing the probability of an earthquake.

S	represents the current level of seismic activity, which could include factors like tremor frequency and ground displacement.
H	denotes a value derived from historical seismic data, such as the frequency and magnitude of past earthquakes in the area.
a, b and c	are coefficients adjusted based on historical data to optimize the model's predictive accuracy.
e	is the base of the natural logarithm.

5.3 Floods

The management and prediction of floods present a complex challenge due to the multifaceted interactions between atmospheric conditions, land use, and human infrastructure. Flooding can result from a variety of factors, including heavy rainfall, snowmelt, dam breaches, or a combination of these. The advent of neuromorphic computing offers a promising advance in the ability to model, predict, and manage flood events with unprecedented precision and speed.

- **Development of neuromorphic computing models for flood prediction**: Neuro-morphic computing models for flood prediction leverage the vast array of environmental and meteorological data available through sensors distributed across river basins, urban areas, and watersheds. These sensors monitor rainfall, soil moisture, river levels, and other critical indicators that contribute to flood risks. Neuromorphic systems analyze this data in real-time, drawing on historical patterns to predict how current conditions might lead to flooding [5].

  What sets neuromorphic models apart is their capacity to simulate complex hydrological processes at a speed and scale not feasible with traditional computing methods. By processing data on-site, these models can quickly identify potential flood events, considering not only immediate rainfall but also antecedent conditions that might exacerbate flooding, such as saturated soils or existing snowpack [6].
- **Application in flood management**: The application of neuromorphic computing in flood management extends to both early warning systems and the operational planning of flood defenses. For early warnings, neuromorphic models can predict the onset of floods with enough lead time to alert communities and initiate evacuation procedures, significantly reducing human and economic losses. This is particularly crucial in urban areas, where the dense population and complex infrastructure systems require more sophisticated modeling to predict flood pathways and impacts accurately [7].

Beyond prediction, neuromorphic computing can also optimize the operation of flood defense infrastructure. This includes the dynamic control of reservoir levels, the activation of flood barriers, and the management of drainage systems to mitigate the impact of floodwater. By predicting flood scenarios in real-time, these systems can make automated decisions on how best to allocate water storage, when to release reservoirs, and where to focus flood mitigation efforts.

• **Comparative advantage**: Compared to traditional flood prediction models, which often rely on generalized hydrological models and static data sets, neuromorphic computing models offer a dynamic and integrated approach to flood prediction and management. Traditional models can struggle with the computational demands of real-time data analysis and the complexity of modeling flood dynamics in heterogeneous landscapes [8].

Neuromorphic models, however, thrive in this environment. Their ability to rapidly process information from a multitude of sources allows for the integration of diverse data sets, from satellite imagery to sensor networks, providing a holistic view of flood risk that is continuously updated. Moreover, the adaptive learning capabilities of neuromorphic computing mean that each flood event, whether predicted accurately or not, serves as a data point for refining future predictions, enhancing the system's accuracy over time.

In essence, neuromorphic computing transforms flood management from a reactive discipline into a proactive one. By leveraging the speed, adaptability, and analytical power of neuromorphic models, communities can anticipate flood events with greater accuracy, safeguarding lives, property, and infrastructure against one of nature's most pervasive threats.

Table 5.1 contrasts traditional hydrological models with those based on neuromorphic computing, across several key parameters important to flood prediction: prediction lead time, accuracy, and computational efficiency. It highlights the benefits neuromorphic computing brings to enhancing the predictive capabilities for flood management.

5.4   Urban Fires

Urban fires pose a significant risk to densely populated areas, with the potential to cause widespread damage to property and infrastructure, as well as loss of life. The challenge of predicting and managing urban fires is compounded by a range of factors, including the urban heat island effect, varying building materials and densities, and the human activities that may ignite fires. Neuromorphic computing offers a novel approach to enhancing our capability to predict, detect, and respond to urban fires more effectively than ever before.

• **Development of neuromorphic computing models for urban fire prediction**: The development of neuromorphic computing models for urban fire prediction involves the deployment of sensor networks across cities to monitor indicators of

Table 5.1 Traditional hydrological models versus neuromorphic computing models in flood prediction

Parameter	Traditional hydrological models	Neuromorphic computing models
Prediction lead time	Often limited to hours or days ahead, depending on the complexity of the model and the availability of data	Extended lead times, potentially days to weeks ahead, due to faster processing of complex data and real-time updates
Accuracy	Variable, can be affected by the lack of real-time data integration and the static nature of model parameters	Significantly improved by the ability to integrate and analyze diverse data sources in real-time, adapting to changing conditions
Computational efficiency	Can be computationally intensive, especially for models covering large geographic areas, leading to delays in output	Enhanced efficiency, with the ability to process large datasets quickly and in parallel, reducing the time to generate predictions
Data integration	Limited by the ability to manually update models with new data sources or types, potentially overlooking critical variables	Dynamically integrates a wide range of data types and sources, continuously updating and refining predictions based on new information
Model adaptability	Static models require manual recalibration to incorporate new data or understandings of hydrological processes	Models learn and adapt automatically, continuously improving their predictive capability with each new data input or event
User accessibility	Often requires specialized knowledge to interpret model outputs and make decisions	Can provide more accessible and understandable predictions, aiding decision-makers in taking timely action

fire risk continuously. These sensors can detect heat, smoke, and changes in air quality that precede fires, feeding data back to neuromorphic processors in real-time. Unlike traditional computing systems, neuromorphic models can analyze this complex, multidimensional data at incredible speeds, identifying patterns and anomalies indicative of emerging fire threats [9].

Neuromorphic models employ advanced algorithms that learn from historical fire data, incorporating variables such as weather conditions, urban layout, and typical human behaviors that contribute to fire risks. By processing current conditions through the lens of these learned patterns, neuromorphic systems can predict the likelihood of urban fires with a high degree of accuracy, potentially even identifying specific areas at greater risk [10].

- **Application in urban fire management**: In managing urban fires, neuromorphic computing can revolutionize both early warning systems and the coordination of firefighting resources. For early warnings, these systems can alert emergency services and the public to emerging fire threats before they escalate, enabling swift evacuation and response efforts that can save lives and minimize damage [11].

Beyond prediction, neuromorphic computing can also enhance the strategic deployment of firefighting resources. By simulating the potential spread of fires in real-time, considering variables such as wind speed, building materials, and the availability of firebreaks, neuromorphic models can guide decision-makers in prioritizing firefighting efforts and deploying resources where they are needed most. This dynamic, data-driven approach ensures that responses to urban fires are both rapid and strategically informed, maximizing the efficiency of firefighting efforts.

• **Comparative advantage**: Compared to traditional models for urban fire prediction and management, which often rely on static risk assessments and reactive response strategies, neuromorphic computing models offer significant advantages. Traditional methods may not account for the real-time interplay of factors that contribute to fire risk, nor can they easily adapt to the unpredictable nature of urban fires once they begin [12].

Neuromorphic computing, with its ability to process vast amounts of data in real-time and learn from each event, provides a more dynamic and adaptive approach. Its predictive capabilities allow for preemptive action, while its analytical power supports more informed and effective firefighting strategies, potentially reducing the impact of urban fires significantly.

In summary, the integration of neuromorphic computing into urban fire management represents a promising advancement in our ability to protect urban populations and their environments from the devastating effects of fires. By harnessing the speed, adaptability, and intelligence of neuromorphic models, cities can enhance their resilience against fire threats, safeguarding both lives and property in increasingly dense urban landscapes.

Table 5.2 outlines the key benefits of incorporating neuromorphic computing into urban fire management strategies. It highlights the technology's potential to transform current practices through enhanced detection speed, predictive accuracy, and the optimization of firefighting resources, offering a new paradigm in managing urban fire incidents.

5.5 Comparative Analysis

The advent of neuromorphic computing brings forth a paradigm shift in the field of natural disaster management, offering sophisticated predictive models that significantly diverge from traditional approaches in terms of efficiency, accuracy, and speed. This comparative analysis delves into the nuanced differences between neuromorphic models and traditional computing methods across various natural disaster scenarios, including earthquakes, floods, and urban fires, to underscore the transformative potential of neuromorphic computing in enhancing disaster preparedness and response.

Table 5.2 Leveraging neuromorphic computing for enhanced urban fire management

Benefit	Description
Enhanced detection speed	Neuromorphic computing systems can process data from sensors in real-time, identifying signs of fire much faster than traditional monitoring systems, enabling quicker activation of emergency responses
Predictive accuracy	By analyzing vast datasets, including historical fire patterns, weather conditions, and urban layouts, neuromorphic models predict fire outbreaks and spread with high precision, allowing for preemptive measures
Dynamic resource optimization	Utilizing real-time data, neuromorphic computing helps in the strategic allocation of firefighting resources, ensuring that personnel, equipment, and water supplies are optimally deployed to manage incidents effectively
Improved situational awareness	Integrating data from various sources (e.g., satellites, drones, ground sensors), neuromorphic systems provide a comprehensive view of the fire situation, enhancing the command center's situational awareness and decision-making capabilities
Automated evacuation routing	Neuromorphic models can simulate multiple evacuation scenarios in seconds, identifying the safest and quickest evacuation routes for civilians, and dynamically updating these routes as the fire or conditions change
Infrastructure protection	Advanced predictive capabilities allow for the identification of critical infrastructure at risk, enabling targeted protection efforts and minimizing potential disruptions to city operations and services
Post-incident analysis and learning	After a fire incident, neuromorphic systems analyze the data to learn from the event, continuously improving prediction models and response strategies for future incidents and enhancing overall urban resilience

I. **Efficiency**
  - *Traditional models*: Conventional disaster prediction models often require substantial computational resources, particularly when processing large datasets or running complex simulations. This computational demand can lead to inefficiencies, particularly in terms of energy consumption and the time required to generate actionable insights [13].
  - *Neuromorphic computing models*: Neuromorphic systems, inspired by the neural structures of the human brain, are inherently more efficient. They are capable of processing information in a parallel and distributed manner, significantly reducing the energy and time needed to analyze vast amounts of data. This efficiency is crucial during disaster events, where rapid response can save lives and minimize damage [14].

II. **Accuracy**
  - *Traditional models*: The accuracy of traditional models largely depends on the quality and granularity of input data. While these models can provide valuable insights, they may struggle with the inherent unpredictability and variability

of natural phenomena, leading to potential inaccuracies in predictions and risk assessments [15].

- *Neuromorphic computing models*: Leveraging the ability to learn and adapt from historical and real-time data, neuromorphic models excel in identifying complex patterns and predicting disaster events with a higher degree of accuracy. Their capacity to process diverse data sources, from satellite imagery to sensor networks, allows for a more nuanced understanding of disaster risks and a reduction in false alarms [16].

III. **Speed**

- *Traditional models*: The speed at which traditional computing models can process data and generate predictions is often constrained by their sequential processing capabilities. In the context of disaster management, this delay can hinder timely decision-making and emergency response efforts [17].
- *Neuromorphic computing models*: One of the standout advantages of neuromorphic computing is its ability to analyze data in real-time, thanks to its parallel processing architecture. This rapid data analysis capability ensures that predictions, alerts, and response strategies can be deployed almost instantaneously, offering a critical advantage in emergencies [18].

IV. **Adaptive Learning**

- *Traditional models*: Traditional disaster management models typically operate on fixed algorithms that do not evolve based on new data. While updates can be made, they often require manual intervention, making it difficult for these models to adapt quickly to changing conditions or incorporate new scientific insights [19].
- *Neuromorphic computing models*: In contrast, neuromorphic models are designed to learn and adapt continuously. They refine their predictions and improve their accuracy over time by incorporating new data and outcomes from past events. This adaptive learning capability ensures that neuromorphic models become more sophisticated and reliable with each disaster event, offering progressively improved guidance for disaster management efforts [20].

Table 5.3 contrasts the adaptive learning capabilities inherent in neuromorphic computing models with the static learning approaches of traditional disaster prediction models. It emphasizes the significant improvements in prediction accuracy and model refinement over time, showcasing the dynamic, evolving nature of neuromorphic computing in disaster management.

5.6 Conclusion

The comparative analysis reveals that neuromorphic computing models offer substantial advantages over traditional models in disaster management, particularly in terms of efficiency, accuracy, speed, and adaptive learning. By harnessing the power of neuromorphic computing, disaster management professionals can significantly

Table 5.3 Adaptive learning in neuromorphic versus static learning in traditional models

Feature	Neuromorphic computing models	Traditional models
Learning approach	Adaptive learning: models evolve based on new data and outcomes, refining predictions continuously	Static learning: models rely on pre-defined algorithms and do not automatically update with new information
Prediction accuracy	Improves over time as the system learns from each event, making future predictions more precise	Remains constant or requires manual updates for improvement, potentially missing critical predictive insights
Response to novel data	Can quickly integrate and learn from novel data sources, enhancing the model's robustness and flexibility	Struggles to incorporate new data types or sources without significant reprogramming or model adjustments
Model refinement	Continuous and automatic, allowing for real-time adjustments to predictive models based on current events and data	Often manual and infrequent, leading to outdated models that may not reflect the latest understanding or data
Scalability	Easily scales with additional data, becoming more accurate and efficient as it processes more information	Scalability is limited; larger datasets can increase computational burden without corresponding increases in accuracy
Customization	Models can dynamically adjust to specific disaster types or geographic areas, providing tailored predictions	Generally broad and less tailored, requiring significant effort to customize for specific scenarios or regions

enhance their ability to predict natural disasters, respond to emergencies more effectively, and ultimately, save lives and protect communities. As technology advances and neuromorphic systems become more integrated into disaster management strategies, we can expect a notable shift toward more proactive and resilient approaches to handling natural disasters.

References

1. M.H. Al Banna, K.A. Taher, M.S. Kaiser, M. Mahmud, M.S. Rahman, A.S. Hosen, G.H. Cho, Application of artificial intelligence in predicting earthquakes: state-of-the-art and future challenges. IEEE Access **8**, 192880–192923 (2020). https://doi.org/10.1109/ACCESS.2020.3029859
2. G. Cremen, C. Galasso, Earthquake early warning: recent advances and perspectives. Earth Sci. Rev. **205**, 103184 (2020). https://doi.org/10.1016/j.earscirev.2020.103184
3. K.W. Campbell, Proposed methodology for estimating the magnitude at which subduction megathrust ground motions and source dimensions exhibit a break in magnitude scaling: example for 79 global subduction zones. Earthq. Spectra **36**(3), 1271–1297 (2020). https://doi.org/10.1177/8755293019899957

4. M.S. Abdalzaher, H.A. Elsayed, M.M. Fouda, M.M. Salim, Employing machine learning and iot for earthquake early warning system in smart cities. Energies **16**(1), 495 (2023). https://doi.org/10.3390/en16010495

5. L. Xu, P. Gober, H.S. Wheater, Y. Kajikawa, Reframing socio-hydrological research to include a social science perspective. J. Hydrol. **563**, 76–83 (2018). https://doi.org/10.1016/j.jhydrol.2018.05.061

6. D. Caviedes-Voullième, J. Fernández-Pato, C. Hinz, Performance assessment of 2D zero-inertia and shallow water models for simulating rainfall-runoff processes. J. Hydrol. **584**, 124663 (2020). https://doi.org/10.1016/j.jhydrol.2020.124663

7. W. Wu, R. Emerton, Q. Duan, A.W. Wood, F. Wetterhall, D.E. Robertson, Ensemble flood forecasting: current status and future opportunities. Wiley Interdiscip. Rev. Water **7**(3), e1432 (2020). https://doi.org/10.1002/wat2.1432

8. D. Wijayarathne, P. Coulibaly, S. Boodoo, D. Sills, Evaluation of radar-gauge merging techniques to be used in operational flood forecasting in urban watersheds. Water **12**(5), 1494 (2020). https://doi.org/10.3390/w12051494

9. W. Nikolakis, E. Roberts, Wildfire governance in a changing world: insights for policy learning and policy transfer. Risk Hazards Crisis Pub. Policy **13**(2), 144–164 (2022). https://doi.org/10.1002/rhc3.12235

10. M. Ambroz, M. Balažovjech, M. Medl'a, K. Mikula, Numerical modeling of wildland surface fire propagation by evolving surface curves. Adv. Comput. Math. **45**, 1067–1103. https://doi.org/10.1007/s10444-018-9650-4

11. V. Agranat, V. Perminov, Mathematical modeling of wildland fire initiation and spread. Environ. Model. Softw. **125**, 104640 (2020). https://doi.org/10.1016/j.envsoft.2020.104640

12. L. Benatti, T. Zanotti, D. Gandolfi, J. Mapelli, F.M. Puglisi, Biologically plausible information propagation in a complementary metal-oxide semiconductor integrate-and-fire artificial neuron circuit with memristive synapses. Nano Futures **7**(2), 025003 (2023). https://doi.org/10.1088/2399-1984/accf53

13. F. Privé, H. Aschard, A. Ziyatdinov, M.G. Blum, Efficient analysis of large-scale genome-wide data with two R packages: bigstatsr and bigsnpr. Bioinformatics **34**(16), 2781–2787 (2018). https://doi.org/10.1093/bioinformatics/bty185

14. A.M. Zyarah, K. Gomez, D. Kudithipudi, Neuromorphic system for spatial and temporal information processing. IEEE Trans. Comput. **69**(8), 1099–1112 (2020). https://doi.org/10.1109/TC.2020.3000183

15. J.F. Torres, D. Hadjout, A. Sebaa, F. Martínez-Álvarez, A. Troncoso, Deep learning for time series forecasting: a survey. Big Data **9**(1), 3–21 (2021). https://doi.org/10.1089/big.2020.0159

16. L. Wang, Y. Zhao, Y. Jinnai, Y. Tian, R. Fonseca, Alphax: exploring neural architectures with deep neural networks and monte carlo tree search (2019). arXiv preprint arXiv:1903.11059. https://doi.org/10.48550/arXiv.1903.11059

17. M. Imran, F. Ofli, D. Caragea, A. Torralba, Using AI and social media multimodal content for disaster response and management: opportunities, challenges, and future directions. Inf. Process. Manage. **57**(5), 102261 (2020). https://doi.org/10.1016/j.ipm.2020.102261

18. A. Mehonic, A. Sebastian, B. Rajendran, O. Simeone, E. Vasilaki, A.J. Kenyon, Memristors—from in-memory computing, deep learning acceleration, and spiking neural networks to the future of neuromorphic and bio-inspired computing. Adv. Intell. Syst. **2**(11), 2000085 (2020). https://doi.org/10.1002/aisy.202000085

19. N. Shahid, T. Rappon, W. Berta, Applications of artificial neural networks in health care organizational decision-making: a scoping review. PLoS ONE **14**(2), e0212356 (2019). https://doi.org/10.1371/journal.pone.0212356

20. Y.X. Hou, Y. Li, Z.C. Zhang, J.Q. Li, D.H. Qi, X.D. Chen, J. Zhang et al., Large-scale and flexible optical synapses for neuromorphic computing and integrated visible information sensing memory processing. ACS Nano **15**(1), 1497–1508 (2020). https://doi.org/10.1021/acsnano.0c08921

Chapter 6
Intelligent Decision-Making Frameworks

This chapter investigates the integration of neuromorphic computing into intelligent decision-making frameworks within disaster management. These frameworks harness the sophisticated capabilities of neuromorphic systems to process vast and varied datasets in real-time, facilitating rapid and precise decision-making during emergencies. The chapter outlines the architectural design of these frameworks and details the crucial role they play in enhancing the efficiency and accuracy of emergency responses. By leveraging real-time data processing, these frameworks improve forecasting, optimize response strategies, and ultimately enhance the safety and resilience of communities against natural disasters. The discussion extends to the specific components of the framework, including sensory data collection, neuromorphic processing, and the critical feedback mechanisms that enable continuous learning and improvement. Through a series of illustrations and comparative analyses, the chapter underscores the transformative potential of neuromorphic computing in reshaping disaster management practices.

6.1 Framework Design

Designing an intelligent decision-making framework that leverages neuromorphic computing involves crafting a system that is not only technologically advanced but also intuitive and adaptable to the needs of disaster management. This design process focuses on creating a harmonious integration of hardware, software, and human expertise to facilitate rapid and accurate decision-making in emergencies.

Figure 6.1 illustrates the architecture of an intelligent decision-making framework, centralized around the Central Processing Unit. It includes "sensory data collection" for raw input acquisition, "neuromorphic processor" for advanced data processing, "integration and analysis" for consolidating and examining data, and

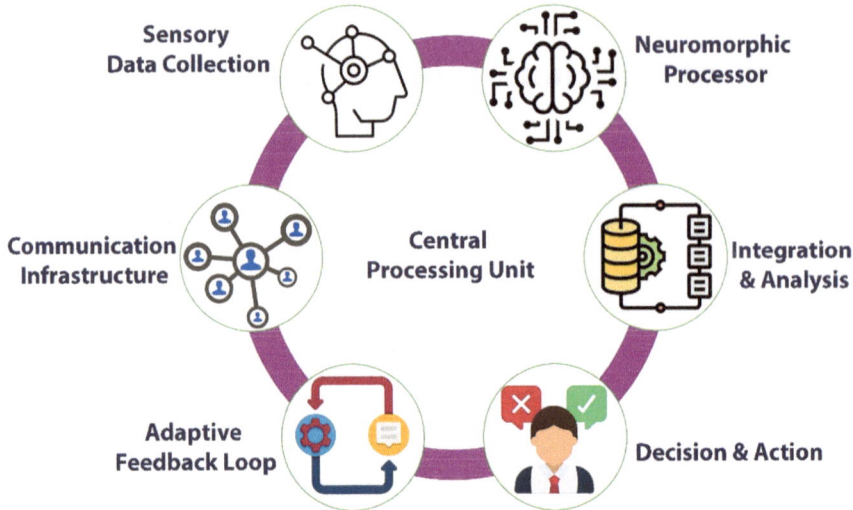

Fig. 6.1 Architecture of an intelligent decision-making framework

"decision and action" for determining the course of action. The system's robustness is further enhanced by an "adaptive feedback loop" for continuous learning, all underpinned by a comprehensive "communication infrastructure" for seamless information exchange.

I. **Core Components of the Framework**

- *Sensory input layer*: The foundational element of this framework is a comprehensive network of sensors, strategically deployed across vulnerable and critical areas. These sensors collect a wide array of data, including but not limited to seismic activity, water levels, weather conditions, and urban infrastructure status. The sensory input layer is designed for robustness and redundancy to ensure continuous data collection even under adverse conditions [1].
- *Neuromorphic processing units*: Central to the framework are the neuromorphic processing units, which represent the brain of the operation. These units are equipped with advanced neuromorphic chips that utilize spiking neural networks (SNNs) to process data in a manner that closely mimics human neural processes. The design of these units focuses on scalability, enabling the integration of additional processing power as data volume grows. Efficiency in energy consumption is also a critical design consideration, ensuring that the system can operate continuously without excessive power requirements [2].
- *Data integration and analysis module*: This module serves as the intermediary between raw data collection and actionable insights. It integrates data from diverse sources, ensuring that the neuromorphic processing units have access to comprehensive, real-time information. Advanced algorithms within

this module analyze the integrated data, identifying patterns, anomalies, and potential threats. The design of this module emphasizes flexibility, allowing for the incorporation of new data sources and analytical methods as they become available [3].

- *Decision and action layer*: The output from the neuromorphic processing units and the data analysis module feeds into the decision and action layer, where critical decisions are made. This layer includes automated systems capable of initiating immediate actions, such as closing floodgates or sending mass alerts, as well as dashboards and interfaces for human decision-makers. The design here is user-centric, ensuring that information is presented clearly and concisely to facilitate quick and informed decisions [4].
- *Feedback and learning mechanism*: A pivotal aspect of the framework's design is its ability to learn and adapt from each event. Incorporating machine learning algorithms, the system analyzes the outcomes of its predictions and actions, continually refining its models and strategies. This feedback mechanism is designed for iterative improvement, ensuring that the system becomes more accurate and effective over time [5].
- *Communication and coordination network*: Recognizing the importance of multiagency coordination in disaster response, the framework includes a communication network that enables seamless information sharing and collaboration among different stakeholders, including emergency services, government agencies, and community organizations. The design of this network prioritizes security and reliability, ensuring that critical information is exchanged efficiently and securely during disasters [6].

II. Design Philosophy

The overarching design philosophy of this intelligent decision-making framework is centered on resilience, adaptability, and user-centricity. It acknowledges the dynamic and complex nature of disaster management, aiming to provide a system that not only responds to current challenges but is also capable of evolving to meet future demands. By integrating neuromorphic computing's unparalleled processing capabilities with a deep understanding of disaster management's needs, the framework is poised to transform how we predict, prepare for, and respond to natural disasters [7].

Table 6.1 outlines the essential components of an intelligent decision-making framework designed for disaster management, detailing their primary functions and how neuromorphic computing plays a crucial role in enhancing these components' effectiveness.

6.2 Real-Time Data Processing

Real-time data processing stands as a critical component in the paradigm of disaster management, especially when underpinned by neuromorphic computing technologies. The essence of real-time data processing in this context lies not just in the speed

Table 6.1 Components and enhanced functions within the intelligent decision-making framework

Component	Primary function	Enhancement through neuromorphic computing
Sensory input layer	Collects real-time data from various sources, including environmental sensors, satellite imagery, and social media	Neuromorphic computing enables faster processing and integration of diverse data types, enhancing the sensitivity and responsiveness of the system
Neuromorphic processing units	Analyzes the collected data using models that mimic the human brain's neural networks, making sense of complex patterns and anomalies	Offers unparalleled speed and efficiency in data processing, allowing for real-time analysis and immediate identification of potential disaster indicators
Data integration and analysis module	Integrates data from multiple sources, providing a unified analysis to identify potential threats and necessary actions	Facilitates the seamless combination of disparate data streams, using adaptive algorithms to improve predictive accuracy and decision-making insights
Decision and action layer	Translates analytical insights into actionable decisions and coordinates the execution of response strategies	Enhances decision-making with predictive analytics, ensuring actions are data-driven and targeted, with automated systems implementing responses without delay
Feedback and learning mechanism	Uses post-event data to refine models and strategies, improving future performance based on lessons learned	Employs machine learning to continuously update and refine prediction models, ensuring the system adapts and evolves with each new piece of information
Communication and coordination network	Facilitates information sharing among all stakeholders, including emergency services, government agencies, and the public	Leverages advanced communication technologies to ensure reliable, secure, and instant sharing of critical information, enhancing coordination and public awareness

of data analysis but in the capacity to interpret, decide, and act upon vast streams of data from myriad sources instantly. This capability is pivotal for forecasting disasters, optimizing responses, and ultimately saving lives and preserving infrastructure.

I. **Importance of Real-Time Data Processing**

- *Immediate threat detection*: The initial moments of a natural disaster are crucial. Real-time data processing enables the immediate detection of threat signatures, such as unusual seismic patterns indicating an earthquake or rapid water level rises forecasting a flood. This rapid detection is essential for activating early warning systems and initiating evacuation orders, significantly reducing potential casualties and property damage [8].

- *Dynamic situation analysis*: As a disaster unfolds, conditions can change rapidly. Real-time data processing allows for the continuous analysis of incoming data, ensuring that response strategies can be adjusted dynamically in response to new information. This agility is critical for navigating the complexities of unfolding emergencies, where delayed information can render response efforts ineffective [9].
- *Enhanced decision support*: In the heat of a disaster response, decision-makers are tasked with making high-stakes decisions quickly. Real-time data processing supports these decisions by providing up-to-the-minute information on the disaster's scope, affected areas, and the status of critical infrastructure, enabling more informed and effective decision-making [10].

II. **Methods of Real-Time Data Processing**

- *Edge computing*: Neuromorphic computing enhances real-time data processing through edge computing, where data analysis occurs close to the data source rather than in a centralized data center. This proximity reduces latency, allowing for quicker detection of patterns and anomalies that signal emerging threats [11].
- *Parallel processing*: Neuromorphic chips excel at parallel processing, simultaneously analyzing data from multiple sources. This approach mirrors the brain's ability to process information, making it inherently suited for the complex, multifaceted data involved in disaster management [12].
- *Adaptive learning algorithms*: Utilizing adaptive learning algorithms, neuromorphic systems can refine their predictive models in real-time, based on continuous data input. This learning capability means that the systems not only react to current conditions but also anticipate future developments based on past and present data trends [13].

III. **Enhancing Real-Time Data Processing**

To maximize the effectiveness of real-time data processing in disaster management, several strategies can be employed:

- *Data prioritization*: Implementing algorithms that can prioritize data based on relevance and urgency ensures that the most critical information is processed first, enhancing response times for the most pressing threats [14].
- *Scalable architecture*: Designing neuromorphic computing systems with scalable architectures ensures that they can accommodate increasing volumes of data without degradation in performance, crucial for adapting to expanding sensor networks and data sources [15].
- *Integration of diverse data sources*: Incorporating data from a variety of sources, including satellite imagery, social media, and traditional sensor networks, can provide a more comprehensive view of disaster scenarios, improving the accuracy of real-time analysis [16].

In summary, real-time data processing facilitated by neuromorphic computing transforms disaster management from a reactive to a proactive discipline. By harnessing the power of edge computing, parallel processing, and adaptive learning in real-time, disaster management frameworks can significantly improve their ability to detect, analyze, and respond to natural disasters, safeguarding communities and minimizing the impact of catastrophic events.

Equation 6.1 showcases the comparative efficiency of data processing in neuromorphic systems against traditional computing frameworks. By factoring in the volume of data processed, the speed of processing, and energy consumption, this formula highlights the superior efficiency (E_N) of neuromorphic computing (N) over traditional methods (T). The efficiency ratio (R) underscores the enhanced performance and energy economy of neuromorphic systems, emphasizing their potential to revolutionize data-intensive applications by offering faster, more energy-efficient processing capabilities. This efficiency is especially critical in real-time applications such as environmental monitoring and disaster response, where rapid and accurate data analysis is paramount.

$$\begin{cases} E_N = \frac{V_N \times S_N}{E_{cN}} \\ E_T = \frac{V_T \times S_T}{E_{cT}} \end{cases} \tag{6.1}$$

where:

E_N and E_T are the efficiencies of neuromorphic and traditional computing systems, respectively.

V_N and V_T represent the volume of data processed by neuromorphic and traditional systems, respectively.

S_N and S_T denote the processing speeds of neuromorphic and traditional systems, respectively.

E_{cN} and E_{cT} indicate the energy consumption of neuromorphic and traditional systems, respectively, for processing the given volume of data.

To directly compare the efficiencies, we can calculate the efficiency ratio (R) as: $R = \frac{E_N}{E_T}$.

6.3   Emergency Response

In the realm of disaster management, the emergency response phase is critical. It is where planning meets action, and the effectiveness of the response can significantly influence the outcome for affected communities. Integrating neuromorphic computing into emergency response frameworks can revolutionize how agencies prepare for, react to, and manage disasters. This approach enhances response times and decision-making accuracy, offering a new level of sophistication in handling emergencies.

I.   **Enhancing Emergency Response Times**

The speed of response following the onset of a disaster is often a decisive factor in minimizing harm. Neuromorphic computing plays a pivotal role here:

- *Automated alert systems*: Implementing systems capable of instantaneously analyzing data to recognize signs of imminent disasters—such as specific seismic patterns indicative of earthquakes or rapidly changing weather data signaling severe storms. These systems can automatically trigger alerts to emergency services and the public, ensuring rapid evacuation and response measures are set in motion [17].
- *Resource allocation*: Neuromorphic systems can quickly assess the scale and scope of a disaster, enabling them to allocate resources efficiently. By analyzing real-time data from multiple sources, these systems can identify the areas in greatest need and ensure that emergency services are directed there first, optimizing the use of limited resources [18].
- *Traffic and evacuation management*: In urban areas, managing evacuation routes to prevent congestion is crucial. Neuromorphic computing can process data from traffic cameras, GPS systems, and other sources in real-time to dynamically adjust traffic signals, open or close routes, and guide the population via mobile alerts, significantly reducing evacuation times [19].

II. Improving Decision-Making Accuracy

The complexity of disaster scenarios requires decisions to be made quickly, often with incomplete information. Neuromorphic computing enhances decision-making accuracy by:

- *Predictive analysis*: Beyond immediate emergency response, neuromorphic systems can predict the evolution of a disaster event, such as the spread of a wildfire or the progression of a flood. This predictive capacity allows decision-makers to anticipate future developments and plan accordingly, rather than merely reacting to current conditions [20].
- *Scenario simulation*: By simulating various response scenarios in real-time, neuromorphic computing can help emergency managers evaluate the potential effectiveness of different strategies, from evacuation routes to resource deployment plans before planning. This approach allows for more informed, strategic choices under pressure [21].
- *Data fusion and interpretation*: In disaster management, data comes from diverse and often siloed sources. Neuromorphic computing excels in integrating and interpreting this data, providing a unified, clear picture of the situation. This comprehensive understanding supports more accurate and timely decision-making during emergencies [22].

In summary, the integration of neuromorphic computing into emergency response frameworks represents a significant advancement in disaster management. By enhancing response times and improving decision-making accuracy, neuromorphic computing enables a more agile, effective response to emergencies, ultimately

Table 6.2 Enhancing emergency response: before and after neuromorphic computing

Emergency response metric	Before neuromorphic computing	After neuromorphic computing integration
Alert times	Often delayed due to manual data analysis and processing, leading to slower public and emergency services notifications	Dramatically reduced through real-time data analysis, enabling quicker public warnings and mobilization of emergency services
Decision accuracy	Limited by human error and the slower processing of complex data, potentially resulting in less optimal decision-making	Significantly improved by leveraging data-driven insights and predictive models, leading to more informed and precise decisions
Resource utilization	Potential misallocation of resources based on incomplete or outdated information, affecting the efficiency of response efforts	Optimized through dynamic, real-time analysis of the situation, ensuring resources are allocated where they are most needed
Response coordination	Challenges in coordinating a timely response among various stakeholders due to communication lags and information silos	Enhanced by improved communication networks, allowing for seamless, instant coordination across all response teams and agencies
Recovery speed	Recovery efforts can be prolonged and less efficient without precise damage assessments and real-time resource management	Accelerated by accurate, real-time data on damage and needs, enabling targeted recovery actions and better allocation of aid
Public safety and trust	Potentially compromised by slower response times and less accurate information, affecting public perception and trust	Strengthened by quicker, more accurate responses and clear, timely communication, enhancing public confidence in disaster management efforts

leading to better outcomes for affected communities and reduced impacts of natural disasters.

Table 6.2 compares key emergency response metrics before and after the integration of neuromorphic computing technologies. It highlights the significant improvements in alert times, decision-making accuracy, and resource utilization, showcasing the transformative effect of neuromorphic computing on emergency response efforts.

References

1. H. Li, K. Ota, M. Dong, Learning IoT in edge: deep learning for the internet of things with edge computing. IEEE Netw. **32**(1), 96–101 (2018). https://doi.org/10.1109/MNET.2018.170 0202

2. Z. Zhou, X. Chen, E. Li, L. Zeng, K. Luo, J. Zhang, Edge intelligence: paving the last mile of artificial intelligence with edge computing. Proc. IEEE **107**(8), 1738–1762 (2019). https://doi.org/10.1109/JPROC.2019.2918951

3. I.G.A. Poornima, B. Paramasivan, Anomaly detection in wireless sensor network using machine learning algorithm. Comput. Commun. **151**, 331–337 (2020). https://doi.org/10.1016/j.comcom.2020.01.005

4. C.M. Vineyard, R. Dellana, J.B. Aimone, F. Rothganger, W.M. Severa, Low-power deep learning inference using the spinnaker neuromorphic platform, in *Proceedings of the 7th Annual Neuro-inspired Computational Elements Workshop* (2019, March), pp. 1–7. https://doi.org/10.1145/3320288.3320300

5. A. Basu, L. Deng, C. Frenkel, X. Zhang, Spiking neural network integrated circuits: a review of trends and future directions, in *2022 IEEE Custom Integrated Circuits Conference (CICC)* (IEEE, 2022, April), pp. 1–8. https://doi.org/10.1109/CICC53496.2022.9772783

6. M. Naeem, T. Jamal, J. Diaz-Martinez, S.A. Butt, N. Montesano, M.I. Tariq, E. De-La-Hoz-Valdiris, Trends and future perspective challenges in big data, in *Advances in Intelligent Data Analysis and Applications: Proceeding of the Sixth Euro-China Conference on Intelligent Data Analysis and Applications, 15–18 October 2019, Arad, Romania* (Springer, Singapore, 2022), pp. 309–325. https://doi.org/10.1007/978-981-16-5036-9_30

7. M. Favaretto, E. De Clercq, C.O. Schneble, B.S. Elger, What is your definition of big data? Researchers' understanding of the phenomenon of the decade. PLoS ONE **15**(2), e0228987 (2020). https://doi.org/10.1371/journal.pone.0228987

8. M. Dabab, M. Freiling, N. Rahman, D. Sagalowicz, A decision model for data mining techniques, in *2018 Portland International Conference on Management of Engineering and Technology (PICMET)* (IEEE, 2018, August), pp. 1–8. https://doi.org/10.23919/PICMET.2018.8481953

9. E. Ismagilova, L. Hughes, Y.K. Dwivedi, K.R. Raman, Smart cities: advances in research—an information systems perspective. Int. J. Inf. Manage. **47**, 88–100 (2019). https://doi.org/10.1016/j.ijinfomgt.2019.01.004

10. K. Ruggeri, S. Alí, M.L. Berge, G. Bertoldo, L.D. Bjørndal, A. Cortijos-Bernabeu, T. Folke et al., Replicating patterns of prospect theory for decision under risk. Nat. Hum. Behav. **4**(6), 622–633 (2020). https://doi.org/10.1038/s41562-020-0886-x

11. B. Adem Esmail, D. Geneletti, Multi-criteria decision analysis for nature conservation: a review of 20 years of applications. Methods Ecol. Evol. **9**(1), 42–53 (2018). https://doi.org/10.1111/2041-210X.12899

12. V. François-Lavet, P. Henderson, R. Islam, M.G. Bellemare, J. Pineau, An introduction to deep reinforcement learning. Found. Trends® Mach. Learn. **11**(3–4), 219–354 (2018). https://doi.org/10.1561/2200000071

13. M.C. Fu, Monte Carlo tree search: a tutorial, in *2018 Winter Simulation Conference (WSC)* (IEEE, 2018, December), pp. 222–236. https://doi.org/10.1109/WSC.2018.8632344

14. Z. Zhang, E. Sejdić, Radiological images and machine learning: trends, perspectives, and prospects. Comput. Biol. Med. **108**, 354–370 (2019). https://doi.org/10.1016/j.compbiomed.2019.02.017

15. A. Lovari, S.A. Bowen, Social media in disaster communication: a case study of strategies, barriers, and ethical implications. J. Public Aff. **20**(1), e1967 (2020). https://doi.org/10.1002/pa.1967

16. T.E. Drabek, Community processes: coordination. Handbook Dis. Res. 521–549 (2018). https://doi.org/10.1007/978-3-319-63254-4_25

17. N. Power, Extreme teams: toward a greater understanding of multiagency teamwork during major emergencies and disasters. Am. Psychol. **73**(4), 478. https://doi.org/10.1037/amp0000248

18. C. Cortinovis, D. Geneletti, Ecosystem services in urban plans: what is there, and what is still needed for better decisions. Land Use Policy **70**, 298–312 (2018). https://doi.org/10.1016/j.landusepol.2017.10.017

19. L. Figueiredo, T. Honiden, A. Schumann, Indicators for resilient cities (2018).https://doi.org/
    10.1787/20737009
20. P. Sharma, A. Joshi, Challenges of using big data for humanitarian relief: lessons from the
    literature. J. Humanit. Logist. Supply Chain Manag. **10**(4), 423–446 (2020). https://doi.org/10.
    1108/JHLSCM-05-2018-0031
21. J.K. Joseph, K.A. Dev, A.P. Pradeepkumar, M. Mohan, Big data analytics and social media
    in disaster management, in *Integrating Disaster Science and Management* (2018, Elsevier),
    pp. 287–294. https://doi.org/10.1016/B978-0-12-812056-9.00016-6
22. T.W. Haase, Uncertainty in crisis management, in *Global Encyclopedia of Public Admin-
    istration, Public Policy, and Governance* (2023, Springer International Publishing, Cham),
    pp. 12957–12961. https://doi.org/10.1007/978-3-030-66252-3_2922

Chapter 7
Challenges and Limitations

As neuromorphic computing advances and finds applications in civil engineering and disaster management, it introduces a host of ethical, legal, and policy challenges. This chapter examines the critical issues surrounding privacy, equity, and accountability that arise from deploying sophisticated computing technologies. It highlights the necessity for rigorous ethical standards, robust legal frameworks, and thoughtful policymaking to manage the integration of these technologies effectively. Through an exploration of these implications, the chapter emphasizes the importance of proactive governance to ensure that the deployment of neuromorphic computing aligns with societal values and legal norms, enhancing disaster resilience while safeguarding individual rights and promoting social justice.

7.1 Technical Challenges

The integration of neuromorphic computing into the domain of civil engineering and disaster management, while promising, is fraught with technical challenges. These challenges stem from the inherent complexities of neuromorphic systems, the vast and varied nature of data involved in disaster management, and the critical need for reliability and resilience in emergency response systems. Addressing these challenges is paramount to harnessing the full potential of neuromorphic computing in this field.

- **Scalability and complexity**: One of the foremost technical hurdles is the scalability of neuromorphic systems. As these systems are designed to mimic the human brain's neural networks, they are inherently complex. Scaling such systems to manage the extensive data streams generated by sensors across vast geographic areas poses significant challenges. Ensuring that these systems can not only handle

A. A. Firoozi, *Neuromorphic Computing*,
SpringerBriefs in Applied Sciences and Technology,
https://doi.org/10.1007/978-3-031-65549-4_7

the volume of data but also process it efficiently and accurately requires advances in hardware design, software algorithms, and system architecture [1].

- **Integration with existing infrastructure**: Another critical challenge lies in integrating neuromorphic computing systems with existing disaster management infrastructure. Many current systems are built on traditional computing architectures that do not easily interface with the neuromorphic models. Achieving seamless integration necessitates developing new protocols and interfaces that can translate between different computing paradigms. This integration must be accomplished without disrupting the ongoing operations of disaster management systems, which are vital for public safety [2].
- **Reliability in extreme conditions**: The operational reliability of neuromorphic computing systems under extreme conditions is another area of concern. Disaster management systems must function flawlessly in the face of natural catastrophes, which can subject them to harsh environmental conditions, power outages, and physical damage. Designing neuromorphic systems that are robust enough to withstand such conditions, including developing fail-safes and redundancies, is a daunting task. Additionally, the systems must be protected against cyber threats, which can escalate during disaster events when vulnerabilities might be more exposed [3].
- **Data privacy and security**: Ensuring data privacy and security in neuromorphic computing systems presents a complex challenge. The vast amounts of data collected for disaster prediction and response include sensitive personal information, critical infrastructure details, and potentially classified information. Protecting this data from breaches requires sophisticated encryption techniques and cybersecurity measures. Moreover, neuromorphic systems, with their distributed processing capabilities, introduce new vulnerabilities that must be addressed to prevent unauthorized access and ensure data integrity [4].
- **Power consumption and environmental impact**: While neuromorphic computing is touted for its efficiency, managing the power consumption of large-scale deployments is a challenge. The environmental impact of operating extensive sensor networks and computing infrastructure, especially in remote or sensitive ecosystems, requires careful consideration. Developing energy-efficient neuromorphic chips and leveraging renewable energy sources for power are avenues being explored to mitigate these impacts [5].

7.2 Addressing the Challenges

Addressing these technical challenges requires a multidisciplinary approach that brings together experts in neuromorphic computing, civil engineering, disaster management, cybersecurity, and environmental science. Collaborative research and development efforts are crucial for overcoming the scalability and integration hurdles, ensuring system reliability, safeguarding data privacy, and minimizing environmental impacts. Advances in material science, encryption technology, and renewable energy will also play key roles in addressing these challenges, paving the way for the successful implementation of neuromorphic computing in disaster management and civil engineering.

Table 7.1 outlines the key technical challenges encountered in the integration of neuromorphic computing within disaster management systems, paired with potential solutions or current strategies aimed at overcoming these obstacles. It provides a clear framework for navigating the complexities of implementing advanced computational technologies in critical emergency response infrastructures.

Table 7.1 Addressing technical challenges in neuromorphic computing integration

Technical challenge	Potential solution or strategy
Scalability and complexity	Implement modular design principles allowing for incremental scaling and integration, facilitating the management of complex systems without compromising performance
Integration with existing infrastructure	Develop interoperable standards and middleware solutions that enable seamless communication and data exchange between neuromorphic systems and legacy computing frameworks
Reliability in extreme conditions	Design robust, fault-tolerant hardware capable of withstanding environmental stressors, complemented by redundant systems to ensure continuous operation during disasters
Data privacy and security	Apply advanced encryption techniques and secure data handling protocols, alongside rigorous access controls, to protect sensitive information from unauthorized access and breaches
Energy consumption	Focus on optimizing neuromorphic chips for low-power consumption and explore the use of renewable energy sources to power sensor networks and computing infrastructures sustainably
Technical expertise gap	Foster education and training programs aimed at developing interdisciplinary expertise, bridging the gap between traditional disaster management practices and neuromorphic computing
Maintenance and upkeep costs	Leverage the inherent efficiency and low maintenance requirements of neuromorphic systems, while also developing cost-effective strategies for system updates and upkeep
Public perception and acceptance	Engage in public outreach and education initiatives to demystify neuromorphic computing, highlighting its benefits for disaster management and addressing concerns transparently

7.3   Ethical and Social Considerations

The deployment of neuromorphic computing in civil engineering and disaster management, while technologically revolutionary, raises several ethical and social considerations that must be carefully navigated. These considerations revolve around the implications of such advanced technology on privacy, equity, and societal trust, highlighting the need for a balanced approach that respects individual rights and promotes social welfare.

- **Privacy concerns**: The use of neuromorphic computing for disaster management involves the collection, processing, and analysis of massive datasets, which can include sensitive personal information. The potential for surveillance and data misuse raises significant privacy concerns. Ensuring that data collection is minimally invasive, and that personal information is anonymized and secured against breaches is crucial. Transparent data policies and robust consent frameworks are essential to maintain public trust and ensure that the benefits of neuromorphic computing do not come at the expense of individual privacy [6].
- **Equity and access**: The benefits of neuromorphic computing in disaster management, such as improved prediction accuracy and response times, should be accessible to all segments of society. However, there is a risk that these benefits could be unevenly distributed, with wealthier communities and nations potentially gaining more access to advanced disaster management technologies. This disparity could exacerbate existing inequalities, particularly in regions prone to natural disasters but lacking the resources to implement such technologies. Addressing this challenge requires a commitment to equity in the deployment of neuromorphic computing solutions, ensuring that vulnerable populations are not left behind [7].
- **Dependence and resilience**: Increasing reliance on neuromorphic computing for disaster prediction and response also brings up concerns about societal overdependence on technology. While these systems can significantly enhance disaster management capabilities, there is a risk that traditional knowledge and resilience strategies could be undervalued or lost. Balancing technological solutions with traditional methods is essential to ensure that communities remain resilient, even when technology fails or is unavailable. Building technological systems that complement rather than replace human judgment and community-based strategies is key to maintaining a holistic approach to disaster resilience [8].
- **Ethical Deployment and Use**: The ethical deployment of neuromorphic computing systems requires careful consideration of their potential impacts on society. This includes assessing the implications of automated decision-making processes, which, while efficient, may not always account for the nuances of human needs and ethical considerations in disaster situations. Developing ethical guidelines and frameworks for the deployment and operation of these systems is crucial to ensure that decisions are made in a manner that prioritizes human welfare and respects ethical principles [9].

- **Engaging with stakeholders**: Addressing ethical and social considerations also involves engaging with a broad range of stakeholders, including affected communities, policymakers, technologists, and ethicists. This engagement should aim to understand diverse perspectives and values, ensuring that the development and deployment of neuromorphic computing in disaster management are guided by a comprehensive understanding of societal needs and ethical considerations. Public awareness and education campaigns can also play a role in demystifying technology and addressing concerns about privacy, equity, and dependence [10].

In summary, the ethical and social considerations associated with the deployment of neuromorphic computing in disaster management underscore the need for a thoughtful and inclusive approach. By addressing privacy concerns, ensuring equitable access, balancing technological dependence with resilience, and adhering to ethical deployment practices, we can harness the benefits of neuromorphic computing to enhance disaster management while respecting individual rights and promoting societal well-being.

Table 7.2 identifies key stakeholders involved in the integration of neuromorphic computing within civil engineering and disaster management systems. It outlines tailored strategies for engaging each group, ensuring their concerns and expectations are addressed, particularly regarding ethical and social considerations.

7.4 Future Perspectives

The integration of neuromorphic computing into civil engineering and disaster management stands at the precipice of a new era, heralding vast possibilities and challenges that will shape the future of these fields. As we venture into this uncharted territory, several potential developments, unresolved questions, and future directions emerge, painting a complex yet promising landscape for neuromorphic computing's role in advancing societal resilience against disasters.

I. **Potential developments**

- *Enhanced sensor technologies*: Future advancements in sensor technologies could greatly expand the types and volumes of data available for neuromorphic processing. Innovations in nanotechnology, bio-sensing, and environmental monitoring could provide more detailed and diverse data streams, enabling neuromorphic systems to generate even more accurate predictions and insights for disaster management [11].
- *Quantum neuromorphic computing*: The potential fusion of neuromorphic computing with quantum computing principles offers exciting prospects. Quantum neuromorphic systems could dramatically increase the speed and capacity of data processing, opening new frontiers in modeling complex systems and scenarios that are currently beyond reach, such as predicting the

Table 7.2 Engaging stakeholders in the adoption of neuromorphic computing

Stakeholder group	Engagement strategy
Policymakers	Collaborate to develop clear guidelines and regulations that support the ethical use of neuromorphic computing, ensuring public safety and data privacy. Engage in dialogue to align technology development with public policy goals
Affected communities	Conduct informational sessions and open forums to educate the public on the benefits and risks of neuromorphic computing, incorporating community feedback into disaster management plans
Tech developers	Encourage the development of user-friendly and accessible technologies, emphasizing the importance of ethical considerations and societal impact in the design phase
Disaster management professionals	Offer training programs and workshops to familiarize these professionals with neuromorphic computing capabilities, focusing on practical applications and ethical deployment in disaster scenarios
Academic and research institutions	Foster collaborative research initiatives that explore the potential of neuromorphic computing in disaster management, including ethical implications and social impact studies
Industry partners	Establish partnerships with technology companies to pilot and refine neuromorphic computing solutions, ensuring they meet the operational and ethical standards required for disaster management
Non-governmental organizations (NGOs)	Work with NGOs to assess the needs of vulnerable populations and integrate neuromorphic computing solutions that enhance disaster resilience and response without exacerbating social inequalities
International organizations	Collaborate on global standards and best practices for the deployment of neuromorphic computing in disaster management, sharing knowledge and resources to address transboundary challenges

outcomes of climate change on specific weather patterns and their impacts on urban areas [12].

- *Autonomous disaster response systems*: Future developments may see the creation of fully autonomous disaster response systems powered by neuromorphic computing. These systems could operate drones, robots, and automated vehicles to perform search and rescue operations, deliver supplies, or even build temporary shelters, all optimized in real-time based on evolving disaster scenarios [13].

II. **Unresolved Questions**

- *Ethical decision-making*: As neuromorphic systems take on more roles in disaster management, questions about the ethics of automated decision-making become more pressing. How do we ensure that these systems make choices that align with human values, especially in life-and-death situations? Developing ethical frameworks that can be integrated into neuromorphic algorithms remains a significant challenge [14].

- *Global standards and protocols*: The lack of global standards for the deployment and operation of neuromorphic computing systems in disaster management raises questions about interoperability, data sharing, and privacy across borders. Establishing international protocols that ensure the safe, ethical, and effective use of this technology is crucial [15].
- *Long-term impacts on employment*: The increasing automation of disaster management tasks through neuromorphic computing may impact employment in this sector. Addressing the potential displacement of jobs and ensuring that workers are retrained for new roles within this evolving landscape are important considerations for the future [16].

III. **Future Directions**

- *Interdisciplinary collaboration*: Advancing neuromorphic computing in civil engineering and disaster management will require increased collaboration across disciplines. Bringing together engineers, computer scientists, environmental scientists, ethicists, and disaster management professionals can foster innovative solutions that address technical, ethical, and social challenges [17].
- *Public engagement and education*: Engaging the public in discussions about neuromorphic computing and its applications in disaster management is essential for building trust and understanding. Public education initiatives can demystify the technology, address concerns, and highlight the benefits of enhanced disaster resilience [18].
- *Policy development and governance*: As neuromorphic computing becomes more integrated into disaster management, developing comprehensive policies and governance structures to oversee its deployment will be vital. These policies should address issues of privacy, equity, and ethical use, ensuring that the technology serves the public good [19].

In conclusion, the future of neuromorphic computing in civil engineering and disaster management is fraught with both challenges and opportunities. By navigating these complex waters with careful consideration of ethical implications, collaborative efforts across fields, and a commitment to societal well-being, we can harness the transformative potential of neuromorphic computing to build a more resilient and prepared society.

Table 7.3 compares the current unresolved questions surrounding the integration of neuromorphic computing within civil engineering and disaster management against potential future directions. It provides insights into how addressing these questions could shape the trajectory of neuromorphic computing applications in enhancing disaster resilience and response.

Table 7.3 Bridging the gap: From unresolved questions to future directions in neuromorphic computing

Unresolved questions	Potential future directions
How can neuromorphic computing be scaled efficiently for global disaster management applications?	Develop scalable neuromorphic systems with cloud integration and decentralized processing capabilities to facilitate global deployment
What are the ethical implications of automated decision-making in disaster scenarios?	Establish ethical frameworks and guidelines for the development and deployment of neuromorphic systems, ensuring they complement human judgment and values
How can data privacy and security be ensured in the widespread collection and analysis of disaster-related data?	Implement cutting-edge encryption and cybersecurity measures and promote transparency in data handling practices to protect sensitive information
What strategies can be employed to bridge the technical expertise gap in neuromorphic computing?	Initiate interdisciplinary educational programs and professional development opportunities focused on neuromorphic computing and its applications in disaster management
How can neuromorphic computing be made accessible and beneficial to communities most vulnerable to disasters?	Design inclusive technologies and deployment strategies that prioritize the needs and capacities of vulnerable populations, ensuring equitable benefits from neuromorphic computing advancements
In what ways can international collaboration be enhanced to leverage neuromorphic computing in disaster management?	Promote international research collaborations and partnerships to share knowledge, technologies, and best practices, fostering a unified global approach to disaster management
What are the long-term impacts of neuromorphic computing on employment within the disaster management sector?	Explore pathways for workforce transition and upskilling, ensuring that professionals are equipped to work alongside advanced technologies in a complementary manner
How can the public trust in neuromorphic computing-enhanced disaster management systems be built and maintained?	Engage in continuous dialogue with the public, offering transparent information on the benefits, risks, and safeguards associated with neuromorphic computing in disaster management

References

1. A. Shrestha, H. Fang, Z. Mei, D.P. Rider, Q. Wu, Q. Qiu, A survey on neuromorphic computing: models and hardware. IEEE Circuits Syst. Mag. **22**(2), 6–35 (2022). https://doi.org/10.1109/MCAS.2022.3166331
2. Z. Kuncic, O. Kavehei, R. Zhu, A. Loeffler, K. Fu, J. Hochstetter, T. Nakayama et al., Neuromorphic information processing with nanowire networks, in *2020 IEEE International Symposium on Circuits and Systems (ISCAS)* (IEEE, 2020, October), pp. 1–5. https://doi.org/10.1109/ISCAS45731.2020.9181034

3. C. Pehle, S. Billaudelle, B. Cramer, J. Kaiser, K. Schreiber, Y. Stradmann, J. Schemmel et al., The BrainScaleS-2 accelerated neuromorphic system with hybrid plasticity. Front. Neurosci. **16**, 795876. https://doi.org/10.3389/fnins.2022.795876

4. B. Rajendran, A. Sebastian, M. Schmuker, N. Srinivasa, E. Eleftheriou, Low-power neuromorphic hardware for signal processing applications: a review of architectural and system-level design approaches. IEEE Signal Process. Mag. **36**(6), 97–110 (2019). https://doi.org/10.1109/MSP.2019.2933719

5. J. Tang, F. Yuan, X. Shen, Z. Wang, M. Rao, Y. He, H. Wu et al., Bridging biological and artificial neural networks with emerging neuromorphic devices: fundamentals, progress, and challenges. Adv. Mater. **31**(49), 1902761. https://doi.org/10.1002/adma.201902761

6. J. Del Valle, J. G. Ramírez, M.J. Rozenberg, I.K. Schuller, Challenges in materials and devices for resistive-switching-based neuromorphic computing. J. Appl. Phys. **124**(21) (2018). https://doi.org/10.1063/1.5047800

7. M. Zhao, B. Gao, J. Tang, H. Qian, H. Wu, Reliability of analog resistive switching memory for neuromorphic computing. Appl. Phys. Rev. **7**(1) (2020). https://doi.org/10.1063/1.5124915

8. S. Choi, J. Yang, G. Wang, Emerging memristive artificial synapses and neurons for energy-efficient neuromorphic computing. Adv. Mater. **32**(51), 2004659 (2020). https://doi.org/10.1002/adma.202004659

9. M. Liu, L. Xia, Y. Wang, K. Chakrabarty, (2018, May). Design of fault-tolerant neuromorphic computing systems. in *2018 IEEE 23rd European Test Symposium* (ETS) (IEEE), pp. 1–9. https://doi.org/10.1109/ETS.2018.8400693

10. N. Hertz, Neurorights–do we need new human rights? A reconsideration of the right to freedom of thought. Neuroethics **16**(1), 5 (2023). https://doi.org/10.1007/s12152-022-09511-0

11. M.U. Hassan, M.H. Rehmani, J. Chen, Differential privacy techniques for cyber physical systems: a survey. IEEE Communications Surveys & Tutorials **22**(1), 746–789 (2019). https://doi.org/10.1109/COMST.2019.2944748

12. D. Liu, Z. Yan, W. Ding, M. Atiquzzaman, A survey on secure data analytics in edge computing. IEEE Internet Things J. **6**(3), 4946–4967 (2019). https://doi.org/10.1109/JIOT.2019.2897619

13. M. Usama, M. Erol-Kantarci, A survey on recent trends and open issues in energy efficiency of 5G. Sensors **19**(14), 3126 (2019). https://doi.org/10.3390/s19143126

14. G. Srinivasan, C. Lee, A. Sengupta, P. Panda, S.S. Sarwar, K. Roy, Training deep spiking neural networks for energy-efficient neuromorphic computing, in *ICASSP 2020–2020 IEEE International Conference on Acoustics, Speech and Signal Processing (ICASSP)* (IEEE, 2020, May), pp. 8549–8553. https://doi.org/10.1109/ICASSP40776.2020.9053914

15. H.B. Barua, K.C. Mondal, Approximate computing: a survey of recent trends—bringing greenness to computing and communication. J. Instit. Eng. (India) Ser. B **100**(6), 619–626 (2019). https://doi.org/10.1007/s40031-019-00418-8

16. I. Ulnicane, W. Knight, T. Leach, B.C. Stahl, W.G. Wanjiku, Framing governance for a contested emerging technology: insights from AI policy. Policy Soc. **40**(2), 158–177 (2021). https://doi.org/10.1080/14494035.2020.1855800

17. A. Tsamados, N. Aggarwal, J. Cowls, J. Morley, H. Roberts, M. Taddeo, L. Floridi, The ethics of algorithms: key problems and solutions. Ethics Gov. Policies Artif. Intell. 97–123 (2021). https://doi.org/10.1007/978-3-030-81907-1_8

18. N. Bostrom, E. Yudkowsky, The ethics of artificial intelligence, in *Artificial intelligence safety and security* (Chapman and Hall/CRC, 2018), pp. 57–69

19. S.M. Mintenig, P.S. Bäuerlein, A.A. Koelmans, S.C. Dekker, A.P. Van Wezel, Closing the gap between small and smaller: towards a framework to analyse nano-and microplastics in aqueous environmental samples. Environ. Sci. Nano **5**(7), 1640–1649 (2018). https://doi.org/10.1039/C8EN00186C

Chapter 8
Case Studies and Real-World Applications

This chapter showcases the practical applications and transformative potential of neuromorphic computing in the realm of civil engineering and disaster management through detailed case studies. These real-world examples illuminate how neuromorphic computing enhances the prediction, mitigation, and management of natural disasters. Covering diverse scenarios such as earthquake early warning systems in Japan, flood management in the Netherlands, and urban fire control in San Francisco, the chapter provides insights into the integration of advanced technologies within existing infrastructures and discusses the significant improvements in disaster response and management efficacy. These case studies not only highlight the capabilities of neuromorphic systems but also outline the ongoing challenges and future directions in leveraging such innovative technologies for enhancing global disaster resilience.

8.1 Neuromorphic Computing in Action

The transformative impact of neuromorphic computing on disaster management and resilience within civil engineering can be best understood through a series of compelling case studies and real-world applications. These instances not only illuminate the practical utility of neuromorphic systems in predicting and mitigating natural disasters but also showcase the integration of these advanced technologies into existing infrastructure, thereby offering insights into their potential to revolutionize the field. Table 8.1 provides a summary of the applications, key technologies, benefits, and challenges associated with the implementation of neuromorphic computing in earthquake early warning systems in Japan, flood prediction and management in the Netherlands, and urban fire spread prediction in San Francisco.

© The Author(s), under exclusive license to Springer Nature Switzerland AG 2024
A. A. Firoozi, *Neuromorphic Computing*,
SpringerBriefs in Applied Sciences and Technology,
https://doi.org/10.1007/978-3-031-65549-4_8

Table 8.1 Comparative overview of neuromorphic computing applications in disaster management

Case study location	Disaster type	Key technologies employed	Primary benefits	Challenges
Japan	Earthquake	Neuromorphic sensors, seismic data analysis	Enhanced early warning times, improved safety measures activation	Scalability, integration with existing systems
The Netherlands	Flood	Sensor networks, hydrodynamic modeling	Accurate flood predictions, proactive flood defense activation	Cost, technical expertise, data privacy
San Francisco, USA	Urban fire	Thermal imaging, sensor networks	Real-time fire spread prediction, effective evacuation planning	Sensor network coverage, prediction reliability

8.2   Earthquake Early Warning Systems

The advent of neuromorphic computing has heralded significant advancements in the realm of earthquake early warning systems (EEWS), particularly exemplified by its application in Japan, a country perennially at risk due to its position along the Pacific "Ring of Fire." Traditional seismic monitoring systems, though effective to a degree, have been limited by their ability to quickly analyze data and disseminate warnings to affected populations and infrastructures. Neuromorphic computing, with its ability to process information in a manner akin to the human brain, has emerged as a transformative solution to these limitations.

- **Implementation in Japan**: Japan's application of neuromorphic computing to its seismic activity monitoring efforts has been groundbreaking. The country has long been a pioneer in deploying sophisticated technologies to mitigate the impacts of earthquakes, and the integration of neuromorphic sensors within its nationwide seismic network represents a significant leap forward. These sensors are designed to rapidly analyze seismic data, specifically identifying the initial p-wave signatures of an earthquake. P-waves, which travel faster than the more destructive s-waves, provide a crucial window for early warning. The neuromorphic sensors excel in this task, processing data at speeds unachievable by conventional computing systems, thus ensuring the timely activation of automated safety measures such as halting trains, shutting down gas lines, and activating public alert systems [1, 2].
- **Technical advancements and societal impacts**: The technical underpinnings of these neuromorphic systems are deeply rooted in their ability to simulate the neural processing patterns of the human brain, allowing for the rapid identification and analysis of seismic signals amidst vast datasets. This is achieved using specialized hardware and algorithms that prioritize energy efficiency and speed, enabling the real-time processing of data without the latency typically associated with traditional computing systems. The societal impacts of this technology are profound, offering not just improved response times but also enhancing the

resilience of critical infrastructure and potentially saving thousands of lives by allowing for preemptive action in the face of imminent earthquake threats.

- **Future directions and challenges**: Looking ahead, the success of neuromorphic computing in Japan's EEWS presents a compelling case for its broader adoption in earthquake-prone regions worldwide. However, challenges remain, particularly in terms of scalability, the integration of neuromorphic systems with existing seismic networks, and the need for international collaboration in data sharing and system interoperability. Additionally, ongoing advancements in neuromorphic technology, including the development of more sophisticated sensors and algorithms, promise to further enhance the accuracy and reliability of earthquake early warnings.

The deployment of neuromorphic computing in Japan's earthquake early warning system stands as a testament to the potential of advanced computing technologies to make a significant impact on disaster management practices. By continuing to refine these systems and address the associated challenges, the global community can make strides toward safeguarding vulnerable populations and infrastructure from the devastating impacts of earthquakes.

8.3 Flood Prediction and Management in the Netherlands

The application of neuromorphic computing in flood prediction and management represents a paradigm shift in how countries vulnerable to flooding, like the Netherlands, prepare for and respond to these natural disasters. The Netherlands, with a significant portion of its land below sea level, has historically been at the forefront of adopting innovative water management solutions. The integration of neuromorphic computing into its flood management systems has pushed the boundaries of what's possible, offering a glimpse into the future of disaster resilience.

- **Neuromorphic computing in Dutch water management**: In leveraging neuromorphic computing, the Netherlands has enhanced its already sophisticated water management infrastructure, which includes an extensive network of dykes, pumps, and barriers. Neuromorphic systems in this context are utilized to process real-time data from sensors distributed across the country's water management infrastructure. These sensors monitor various parameters, such as water levels, flow rates, and weather conditions, feeding a continuous stream of data into neuromorphic computing systems. These systems, designed to mimic the neural processes of the human brain, analyze the data with remarkable speed and efficiency, enabling the prediction of flood events with unprecedented accuracy [3, 4].
- **Impact on flood prediction and response**: The primary advantage of neuromorphic computing in flood management lies in its ability to rapidly simulate and predict complex hydrodynamic models that traditional computing systems process much slower. This speed is crucial in emergency situations where every second counts. For example, the ability to quickly predict the impact of heavy rainfall

on river levels allows for timely decisions on whether to strengthen dykes, raise flood barriers, or activate pumps to divert water flow. This proactive approach to flood defense, enabled by neuromorphic computing, minimizes the potential for damage and loss of life.

Moreover, the accuracy of these predictions means that emergency responses can be more finely tuned to the severity of the threat, reducing unnecessary evacuations and enabling more targeted deployment of resources. This not only conserves valuable emergency management resources but also reduces the social and economic impact of flood events.

- **Challenges and future directions**: While the integration of neuromorphic computing into flood management systems offers significant benefits, there are challenges to its wider adoption. These include the high cost of developing and maintaining such advanced systems, the need for technical expertise to operate them, and the importance of ensuring the privacy and security of the data they process. Additionally, as climate change increases the frequency and severity of flooding events, there will be a growing need to continuously adapt and improve these systems to keep pace with changing environmental conditions.

Looking forward, the success of the Netherlands in applying neuromorphic computing to flood management is likely to inspire other nations to explore similar technologies for their own disaster management needs. Further advancements in sensor technology and computing power are expected to enhance the capabilities of these systems, potentially incorporating predictive analytics for other types of natural disasters as well. As neuromorphic computing continues to evolve, its role in safeguarding communities from the impacts of climate change and natural disasters will undoubtedly expand.

8.4 Urban Fire Spread Prediction in San Francisco

The challenge of managing and mitigating urban fires, particularly in areas prone to wildfires, requires innovative solutions that can predict and respond to fire spread in real-time. San Francisco, situated in a region frequently threatened by wildfires, has become a testing ground for the application of neuromorphic computing in enhancing urban fire spread prediction and response systems. The city's pilot project in this domain exemplifies how cutting-edge technology can be leveraged to protect communities and infrastructure from the devastating impact of fires.

- **Implementation of neuromorphic computing for fire prediction**: San Francisco's approach to wildfire management integrates neuromorphic computing with a network of sensors and satellite imagery to create a dynamic, real-time predictive model of fire spread. The project utilizes neuromorphic chips, which process information in a manner akin to the human brain, enabling the rapid analysis of vast datasets from diverse sources. This includes thermal imaging from satellites, data from ground-based sensors monitoring temperature, humidity, and wind speed, as well as historical data on previous wildfires [5, 6].

The neuromorphic system's ability to quickly process and analyze this data allows for the prediction of a fire's path and speed with a degree of accuracy and speed previously unattainable. By understanding how a fire is likely to spread, emergency response teams can more effectively deploy resources, issue evacuation orders, and implement containment strategies, significantly mitigating the impact on affected communities.

- **Advantages and societal impacts**: The primary advantage of using neuromorphic computing for urban fire spread prediction lies in its potential to save lives and reduce property damage. In scenarios where every moment counts, the system's rapid analysis and prediction capabilities allow for quicker decision-making and response. Furthermore, the ability to accurately predict the spread of fires enables more precise evacuation orders, reducing panic and confusion among the population.

  Additionally, this technology can play a crucial role in resource allocation. By predicting the most likely path of a fire, emergency services can optimize the deployment of firefighting resources, ensuring that they are concentrated where they are most needed, thereby enhancing the efficiency of response efforts.

- **Challenges and future directions**: Despite its promising potential, the implementation of neuromorphic computing in fire spread prediction faces several challenges. These include the need for extensive sensor networks to provide the necessary data, the integration of this technology with existing emergency response frameworks and ensuring the reliability and accuracy of the predictions under rapidly changing conditions.

Looking ahead, continuous advancements in neuromorphic computing, sensor technology, and machine learning algorithms are expected to further enhance the capabilities of urban fire spread prediction systems. There is also potential for this technology to be applied in other contexts, such as in the management of chemical spills or in urban planning to design cities that are more resilient to the threat of wildfires.

As the technology matures and becomes more integrated into disaster management strategies, it holds the promise of transforming how cities around the world prepare for and respond to the threat of urban fires, ultimately making communities safer and more resilient.

References

1. M.S. Abdalzaher, H.A. Elsayed, M.M. Fouda, M.M. Salim, Employing machine learning and iot for earthquake early warning system in smart cities. Energies **16**(1), 495 (2023). https://doi.org/10.3390/en16010495

2. M.S. Abdalzaher, M.S. Soliman, S.M. El-Hady, A. Benslimane, M. Elwekeil, A deep learning model for earthquake parameters observation in IoT system-based earthquake early warning. IEEE Internet Things J. **9**(11), 8412–8424 (2021). https://doi.org/10.1109/JIOT.2021.3114420

3. C. Chen, J. Jiang, Y. Zhou, N. Lv, X. Liang, S. Wan, An edge intelligence empowered flooding process prediction using Internet of things in smart city. J. Parall. Distrib. Comput. **165**, 66–78 (2022). https://doi.org/10.1016/j.jpdc.2022.03.010

4. P. Mitra, R. Ray, R. Chatterjee, R. Basu, P. Saha, S. Raha, S. Saha et al., Flood forecasting using internet of things and artificial neural networks, in *2016 IEEE 7th Annual Information Technology, Electronics and Mobile Communication Conference (IEMCON)* (IEEE, 2016, October), pp. 1–5. https://doi.org/10.1109/IEMCON.2016.7746363

5. Y. Zhang, P. Geng, C.B. Sivaparthipan, B.A. Muthu, Big data and artificial intelligence based early risk warning system of fire hazard for smart cities. Sustain. Energy Technol. Assess. **45**, 100986 (2021). https://doi.org/10.1016/j.seta.2020.100986

6. H. Kaur, S.K. Sood, Energy-efficient IoT-fog-cloud architectural paradigm for real-time wildfire prediction and forecasting. IEEE Syst. J. **14**(2), 2003–2011 (2019). https://doi.org/10.1109/JSYST.2019.2923635

Chapter 9
Ethical, Legal, and Policy Implications

This chapter delves into the complex ethical, legal, and policy challenges introduced by the integration of neuromorphic computing into civil engineering and disaster management. It discusses the pivotal need for rigorous standards and frameworks to navigate issues of privacy, data protection, and equitable access. The chapter highlights how neuromorphic computing's capabilities in real-time data processing and decision-making in emergency scenarios raise significant concerns about accountability, transparency, and public trust. By examining these implications, the chapter underscores the necessity of a proactive approach, involving diverse stakeholders to ensure that technological advancements align with societal values and legal standards, ultimately fostering a secure and equitable deployment of neuromorphic computing in critical infrastructures.

9.1 Navigating the Complexities

As neuromorphic computing technologies continue to evolve and find applications within civil engineering and disaster management, they bring not only innovative solutions but also a host of ethical, legal, and policy implications that necessitate careful consideration. This section explores these implications, aiming to provide a comprehensive overview that underscores the importance of addressing these challenges proactively. Table 9.1 outlines the key policy implications associated with the deployment of neuromorphic computing technologies in the realm of disaster management and civil engineering, highlighting areas such as standardization, public engagement, and resource allocation. This summary aims to guide policymakers, researchers, and practitioners in navigating the complexities and opportunities presented by these advanced technologies.

© The Author(s), under exclusive license to Springer Nature Switzerland AG 2024　　　83
A. A. Firoozi, *Neuromorphic Computing*,
SpringerBriefs in Applied Sciences and Technology,
https://doi.org/10.1007/978-3-031-65549-4_9

Table 9.1 Policy implications of neuromorphic computing in disaster management

Policy area	Description	Recommended actions
Standardization and interoperability	The need for international standards and protocols to ensure neuromorphic computing systems can operate seamlessly across borders, enhancing global disaster resilience	Develop and adopt international standards; promote interoperability through cross-border collaborations
Public engagement	Balancing technological innovation with the necessity of public discourse to address ethical, privacy, and equity concerns	Establish channels for public input; ensure transparency and conduct impact assessments
Resource allocation	Strategic investment in research and development of neuromorphic computing to accelerate innovation and deployment in disaster management	Increase funding for basic and applied research; support public–private partnerships; prioritize translational initiatives

9.2 Ethical Considerations

Privacy and data protection emerge as paramount concerns when integrating neuromorphic computing technologies with disaster management systems. These advanced technologies often rely on extensive data collection through sensors, satellites, and other devices to predict and mitigate the effects of natural disasters. While this data collection is crucial for the functionality and effectiveness of such systems, it also poses significant risks to individual privacy. The collection of sensitive information, including the tracking of people's movements or the monitoring of private properties, necessitates stringent measures to ensure that data is handled with the utmost respect for privacy. Safeguards must be established to protect this information from unauthorized access or misuse, ensuring that the benefits of technology do not come at the expense of individual rights.

Equity and accessibility present another ethical dimension. The deployment of neuromorphic computing in disaster management holds immense potential to save lives and reduce property damage. However, this potential can only be fully realized if these benefits are accessible to all segments of society, including those who are most vulnerable. There is a risk that the deployment of these technologies could disproportionately favor affluent or technologically advanced communities, leaving behind marginalized groups who may be more susceptible to the impacts of natural disasters. Ensuring equitable access to the protections afforded by neuromorphic computing is essential, requiring deliberate strategies to deploy these technologies in a manner that includes and benefits all communities.

Accountability and transparency in the use and deployment of neuromorphic computing systems are critical for maintaining public trust. As these systems play increasingly significant roles in critical decision-making processes, from issuing early warnings to guiding evacuation efforts, it becomes essential to establish clear accountability for their outcomes. This includes delineating who is responsible for the accuracy of the predictions and the decisions made based on those predictions. Moreover, the processes by which these systems operate must be transparent, allowing for public scrutiny and understanding. Such transparency is crucial not only for building confidence in these technologies but also for enabling ongoing evaluation and improvement of their effectiveness.

In addressing these ethical considerations, stakeholders from technology developers to policymakers must engage in continuous dialogue. This dialogue should aim to balance the innovative potential of neuromorphic computing with the imperative to protect individual rights, ensure equitable access, and maintain accountability. By navigating these ethical challenges thoughtfully, the integration of neuromorphic computing into disaster management can achieve its full potential, enhancing resilience and safety for all members of society.

9.3 Legal Implications

Delving into the legal implications associated with the implementation of neuromorphic computing in the fields of civil engineering and disaster management reveals a complex landscape of regulatory compliance, liability, and responsibility. These technologies, while promising in enhancing disaster preparedness and response, intersect with existing legal frameworks that govern privacy, data protection, cybersecurity, and emergency management practices. Navigating this intersection requires a nuanced understanding of both the capabilities of neuromorphic computing and the legal principles that aim to safeguard public interest.

Regulatory compliance stands as a significant challenge in the deployment of neuromorphic computing systems. As these technologies gather, analyze, and act upon vast amounts of data, they must do so within the bounds of laws designed to protect individual privacy and ensure data security. This is no small feat, given the rapid pace of technological advancement and the often-slower evolution of legal frameworks. For example, in the European Union, the General Data Protection Regulation (GDPR) sets stringent requirements for data handling and privacy protection. Neuromorphic computing systems deployed in disaster management must be designed and operated in ways that comply with such regulations, a task that requires ongoing vigilance and adaptation as both technology and legal standards continue to evolve.

Liability and responsibility in the context of neuromorphic computing involve determining who is accountable when things go wrong. If a neuromorphic-based early warning system fails to predict a natural disaster accurately, resulting in loss of life or significant property damage, the question of who bears the legal responsibility becomes critically important. Is it the developers of the neuromorphic technology, the operators of the warning system, or the governmental entities that failed to respond adequately to the warnings issued? These questions are complicated by the fact that neuromorphic computing systems, with their ability to learn and adapt, may not always behave in predictable ways. Establishing clear lines of liability is crucial for ensuring that victims of such failures have recourse to justice, while also providing a framework for accountability that can guide the development and deployment of these technologies.

In addition to these challenges, the international nature of disaster management—where natural disasters do not respect national boundaries—calls for a harmonized legal approach to the deployment of neuromorphic computing systems. This necessitates international cooperation to develop legal frameworks and standards that can accommodate the cross-border nature of both technological deployment and natural disasters. Such cooperation would ensure that neuromorphic computing technologies can be utilized effectively and responsibly across jurisdictions, maximizing their potential to protect and aid populations in the face of disaster.

Addressing the legal implications of neuromorphic computing in disaster management thus requires a collaborative effort among technologists, legal experts, policymakers, and international bodies. By fostering dialogue and cooperation across these disciplines, stakeholders can ensure that the deployment of neuromorphic computing technologies not only advances disaster management capabilities but does so in a manner that respects legal principles and protects the rights and safety of individuals and communities.

9.4 Policy Implications

The integration of neuromorphic computing into disaster management and civil engineering introduces significant policy implications that necessitate thoughtful consideration and action. These implications encompass the need for standardization and interoperability, the imperative to balance innovation with public engagement, and the strategic allocation of resources toward research and development. Addressing these policy implications is essential for maximizing the benefits of neuromorphic computing technologies while ensuring they align with societal goals and values.

The development of international standards and protocols is crucial for the effective deployment of neuromorphic computing systems across the globe. Natural disasters, by their very nature, do not adhere to geopolitical boundaries, making the interoperability of disaster management technologies a matter of international concern. Policies aimed at promoting standardization can facilitate the sharing of data and resources, enhance the efficiency of cross-border disaster response efforts, and ensure that innovations in neuromorphic computing can be rapidly and effectively deployed in disaster-prone regions worldwide. Such standardization also aids in addressing compatibility issues, enabling diverse systems to communicate and work together seamlessly, thereby bolstering global disaster resilience.

   Balancing the drive for technological innovation with the need for public engagement presents another policy challenge. Neuromorphic computing, with its potential
to revolutionize disaster management, also raises questions about privacy, equity,
and the ethical use of technology. Policymakers must navigate these concerns by
fostering an environment where technological advancement is paired with robust
public discourse. This involves creating channels for public input into policy decisions, ensuring transparency in the deployment of neuromorphic computing technologies, and conducting impact assessments to understand the social and ethical
ramifications of these systems. Engaging the public in this way not only enhances the
legitimacy and acceptability of neuromorphic computing initiatives but also ensures
that they are deployed in ways that reflect societal values and priorities.

   The strategic allocation of resources toward the research and development of
neuromorphic computing technologies is also a critical policy consideration. Given
the potential of these technologies to enhance disaster preparedness and response,
governments and international organizations must consider investments in neuromorphic computing a priority within their science and technology portfolios. This
includes funding for basic research, support for public–private partnerships, and
initiatives aimed at translating research findings into practical applications. Such
investments can accelerate the development of neuromorphic computing solutions
that are both effective and ethically responsible, ensuring that the benefits of these
technologies are realized as quickly and equitably as possible.

   In conclusion, the policy implications of integrating neuromorphic computing
into disaster management and civil engineering are profound and wide-ranging. By
addressing the need for standardization and interoperability, fostering public engagement, and strategically allocating resources toward research and development, policymakers can guide the development of neuromorphic computing technologies in a
direction that maximizes their societal benefit. In doing so, they can ensure that these
advanced technologies contribute to building a safer, more resilient world.

Chapter 10
Toward a Resilient Future: The Path Ahead

As we conclude this exploration of neuromorphic computing's transformative potential within the realms of civil engineering and disaster management, it is evident that we stand on the cusp of a technological revolution. This book has traversed the intricate landscape of neuromorphic computing, from its theoretical underpinnings and practical applications to the challenges and ethical considerations it presents. The journey through advanced predictive models for natural disasters and the design of intelligent decision-making frameworks underscores a future where technology and human expertise converge to mitigate the impacts of catastrophic events.

The advent of neuromorphic computing, with its unparalleled efficiency, accuracy, and speed, heralds a new era of disaster resilience. Through the lens of case studies on earthquakes, floods, and urban fires, we have seen the potential of neuromorphic systems to enhance prediction capabilities, improve emergency response times, and ensure more informed decision-making processes. These advancements promise not only to save lives and protect infrastructure but also to fundamentally alter our approach to disaster preparedness and response.

However, this technological journey is not without its challenges. Technical hurdles, such as scalability and integration with existing systems, alongside ethical and social considerations, including privacy concerns and equity in access, present complex obstacles that must be navigated with care. Addressing these challenges requires a multidisciplinary effort, bringing together expertise from across the spectrum of science, engineering, ethics, and policymaking to ensure that the deployment of neuromorphic computing technologies aligns with societal values and priorities.

Looking ahead, the future of neuromorphic computing in civil engineering and disaster management is brimming with possibilities. Continued advancements in sensor technologies, coupled with innovations in quantum neuromorphic systems and autonomous response mechanisms, promise to further enhance our capabilities. Yet, as we chart this path forward, a sustained focus on ethical deployment, global collaboration, and public engagement will be essential to harness the full potential of these technologies for the betterment of society.

© The Author(s), under exclusive license to Springer Nature Switzerland AG 2024   89
A. A. Firoozi, *Neuromorphic Computing*,
SpringerBriefs in Applied Sciences and Technology,
https://doi.org/10.1007/978-3-031-65549-4_10

In conclusion, the integration of neuromorphic computing into disaster management represents a paradigm shift toward more resilient and adaptive infrastructure systems. By embracing this technological revolution, we can envision a future where communities are better prepared to face the challenges posed by natural disasters, safeguarding the well-being of individuals and the built environment against the unpredictable forces of nature. As we move forward, the journey of neuromorphic computing is not merely about advancing technology but about redefining our collective resilience in the face of adversity.